GLOWFACE

Glowface

What We Are Losing to Screens and How We Can Take It Back

DAVID TALBOT

Copyright © 2023 by David Talbot

All rights reserved. No part of this book may be reproduced in any manner whatsoever without written permission except in the case of brief quotations embodied in critical articles and reviews.

Second Printing, 2023

It takes many hours to make what you want to make. The hours don't suddenly appear. You have to steal them from comfort.

Derek Sivers

This book is dedicated to everyone out there stealing hours from comfort.

Contents

Dedication v

INTRODUCTION

1	The Loss of Talents	6
2	The Loss of Depth	33
3	The Loss of Careers	62
4	The Loss of Leadership	81
5	The Loss of Education	115
6	The Loss of Memory	142
7	The Loss of Leisure Time	163
8	The Loss of Mental Health	185
9	Afterword: The Loss of Our Humanity	213

REFERENCES

Introduction

I first heard the term 'glowface' at a party in my hometown of Gymea, a suburb in Sydney's south. It was the middle of winter, and the temperature outside was only about twelve degrees Celsius. When it comes to the weather, Sydneysiders get incredibly dramatic. If the temperature goes into the thirties, it's as if we have been transported to the Sahara – it's simply too hot to exist. If it drops into single digits, the whole city starts to freak out – gloves, beanies and scarfs, and even thermal undergarments, get pulled out of the closet. So to overcome the 'freezing' cold, my friends had lit a fire for us to sit around.

The party was one of those gatherings where you know a few people, but you don't really connect with anyone: the kind of event where the conversation typically remains very shallow. I got to the point where asking what others did for work became somewhat tiresome, and so I gave up asking and started to listen in on other people's conversations.

'You're being a bit of a glowface there,' my friend said to someone sitting around the fire. I had never heard the term, so I asked him what it meant. He explained that he likes to go camping and, when camping, often needs to light a fire. He then talked me through a typical scene: a dark winter's night with a cloudy sky, the only source of light the fire itself – until someone pulls out their phone and starts to become a glowface.

When the only available light is the orange light emanating from a fire, the blue light of a phone really stands out. The result is, the person replying to a text or scrolling through social media looks as if their face is glowing, and so by engaging with their phone, they have become a glowface.

A glowface for our purposes is anyone who is addicted to a digital screen of any kind: anyone whose face is constantly illuminated by the glow of their mobile phone, computer, television, tablet device or even their smart watch. So, let's be honest, if that's the definition of a glowface, we can all be branded one.

What stood out for me about my introduction to the word 'glowface' was the context: we were all sitting around a fire. This is because our relationship to the screens in our lives, to technology in general, has quite a parallel to our relationship with fire. When we are in control of fire, it's an incredible thing. Fire is the reason we humans have become the species we are today. Fire provides us with the ability to cook food that we wouldn't otherwise be able to consume. It provides us with protection from predators or enemies. It's an irreplaceable source of heat on cold Sydney nights or on a winter camp out in the country. It has allowed us to create industries out of melting and moulding metals such steel and aluminium. Once we harnessed fire, we gained the edge over other species and developed into the people we are today. But that's only when we are in control of fire.

When fire is out of our control, it can devastate entire cities in a matter of hours. In these situations, fire doesn't care who you are, where you are from or what you are trying to do to stop it. It can overwhelm you and take your life.

When we are in control of fire, it gives to us. When fire is in control, it takes from us. Technology is no different. When we are in control of technology, it can be used to change the world for the better. Computers have changed how everything works from agriculture, to medicine, to manufacturing. It has revolutionised how we do business and has facilitated globalisation.

Technology also deeply affects our personal lives. It allows us to stay in touch with friends and family and helps us reconnect with people from our past. It has simplified and changed experiences like shopping, ordering food, banking, and has made it easier and safer to hail a ride. A search on the internet can answer most questions within a few moments, and YouTube can teach us how to do almost anything.

Like fire, when we are in control of technology it adds a great deal to our lives – it gives to us. But when it is out of our control, it can take over our life.

Glowface is the story of how the technology and screens in our lives can take from us, steal from us, often without us even knowing. It is a story that challenges the notion that technology is exclusively a force for progress and improvement. It is the story of the consequences of being a glowface.

I want to make it clear at the outset, I'm not against technology. Technology is allowing me to write these very words. It's technology that will allow me to share my thoughts with people like you who are in the process of trying to understand more about what modern technology is really doing to us. But it's this technology that is consuming us every day of our lives and impacting our ability to experience the world around us. The goal of *Glowface* is to shed light on this issue and start a meaningful discussion around how we can change our relationship with technology and take back what we have lost.

Glowface will examine how technology can deprive people of the opportunity to develop their innate talents.

It will show how screens are designed to distract us and how this distraction is shaping our brains in a dangerous way.

It will examine the difficulties faced by those in the corporate world who are forced to work with screens on a daily basis. It will consider the impact of screens and distraction on our leaders. And it will challenge the notion that students learn better with a computer or tablet device.

It will consider how one of the most important aspects of who we are, our memory, is slowly being replaced by technology, and how technology has changed what were once relaxing hobbies and removed their true value.

Finally, *Glowface* will explore the impact of screens on our mental health before turning to the future to consider what dangers it might hold and what this might mean for humanity as a whole. This is the story of what is taken from you, what is stolen from you, when you become a glowface.

Chapter 1

The Loss of Talents

Things are in the saddle/And ride mankind.
Ralph Waldo Emerson

In April 2019, my wife and I decided to take a last-minute babymoon to Anaheim, California, to make all of our kid-afornia dreams come true. We stayed at a hotel only fifteen-minutes' walk from Disneyland and frequently enjoyed stopping at one or other of the many restaurants along the way. We decided one morning to stop in at an IHOP pancake restaurant on Harbour Boulevard, just around the corner from Disneyland. Being from Australia, my wife and I aren't used to the high level of service you generally get at restaurants in America. In Australia, meals often take a while to come out of the kitchen, particularly at cafés. There's a lot of work involved in smashing an avocado, putting it on toast and preparing a single origin pour-over coffee! But, in America, California in particular, service at restaurants is fantastic. The staff are always nice to you, and your food comes out before you even have a chance to put your menu down. As much as I'm sure making a living off tips is difficult for hospitality workers in America, it certainly incentivises them to take care of your entire dining experience.

But for some reason, on this particular morning at the IHOP on Harbour Boulevard, we were forgotten. We looked around the restaurant and started – as you do when you think you've been forgotten at a restaurant – taking stock of who was seated before us and who was seated after us. We soon decided that we definitely put our order in before the table in the corner. And when they had finished their choc-chip pancakes and were asking for the bill, and we were still sitting at our table downing our third cup of water and fighting the urge to go to the bathroom in case our food decided to come, the realisation that something had gone wrong set in. We finally enquired with our waiter, and in the end we got our food. So, crisis averted.

We've all been there – sitting in a restaurant feeling all but forgotten by the wait staff. Often when this happens, we fall into the trap of going on our phones. It's a habit that is reinforced because it's comforting. It avoids the awkwardness and boredom that confronts us when we are waiting. But my wife and I make a conscious effort to be more concerned with being where we are, rather than being where our phones take us. We are quite happy to sit in silence. And if one or other of us needs to go on their phone for some reason, we try our best to share the screen by reaching across the table. We aren't perfect, however, and do sometimes fall victim to the allure of our phones.

It's a difficult temptation to fight. Screens at the table have become an increasingly familiar presence for all ages, but it's children who seem most engrossed in their devices around mealtimes. I recall the day at IHOP specifically because, as we surveyed the room taking note of everyone who was enjoying their pancakes while we were still waiting for ours, we came across the all too familiar sight of a family sitting down for a meal, with their young son being entertained by a screen, in this case an iPad.

The boy was probably no more than ten years old. He was wearing large headphones and was completely engrossed in the content on his screen. I think he barely moved when the pancakes were placed delicately in front of him so as to not disturb his viewing experience.

I remember a similar instance, when we were at a fine dining restaurant in Noosa, a resort town on the east coast of Australia. This time it was a family of four, who were seated at a table not too far from our own – two young boys and two adults. Both young boys, again no more than ten years old, were attached to their devices. They were both thoroughly engrossed and very much wanted to be left alone. Mum was also on her phone, which left Dad without a face to engage with. I don't think he was the biological father of these children because they addressed him by his first name. I mention this because, even from our table, I could almost feel his desire to engage with the boys and win them over. He tried multiple times to spark conversation by tapping one of the boys on their shoulder so they could remove their headphones. But any attempt by this man was met with a violent swipe of the hand and a loud retort. At one point he even asked one of the young boys if they wanted to get ice cream after dinner. Normally, this question guarantees success. What child doesn't want ice cream for dessert? But the boy simply responded with, 'Leave me alone!'

I couldn't help but feel terribly sorry for this guy. He was trying to reach out to these two young boys, but was being rebuffed. Mum wasn't being particularly helpful; she was on her phone throughout this whole episode. That left 'Dad', by himself, fighting the impossible battle of trying to win the attention of these two boys who were captivated by their devices.

It's a scene that is all too familiar – a family sitting down for a meal all but lost to the glow of their screens. If you've picked up this book, chances are you've witnessed that scene too. Maybe you've even been a part of such a scene. And there is probably something telling you that however normal this has now become, however socially acceptable it is, that it isn't right.

These restaurant scenes are a reflection of the broader issues facing children these days. Screens are permeating every aspect of their existence, often without resistance from those responsible for their welfare. Researchers are just starting to scratch the surface when it comes to understanding the cost of this constant exposure. They've studied a variety of areas, and we'll explore many of their findings throughout this book. But an area that is rarely considered is the impact of screens on children's natural talents, how the glow of the screen is a barrier to the development of their abilities. To explore this idea, we will start by considering the impact of screens on the talents of a generation of people which grew up with them – mine.

Guild Wars

I think the impact of screens in the lives of children resonates with me because my childhood, teenage years and subsequent adulthood have been greatly impacted by my relationship with screens. The motivation for writing *Glowface* comes from my experience with screens and how they have uniquely shaped my life.

Back in 1995, when I was just three years old, one of the many activities I enjoyed was maths. I loved maths, which is quite an odd thing to say about a three-year-old, but it turns out I had a natural talent for maths, and it showed itself at a very young age. My mother tells me I would randomly come and tell her things to do with numbers – tell her my multiplication tables without her having taught them to me, tell her how many bricks were on the living room wall and tell her that given my brother was two years older than me, I would be twenty-seven when my brother was twenty-nine. I once analysed a digital clock and worked out that the only numbers that appear simultaneously (the same digit appearing in three positions at the same time) are the numbers one to five because six to nine only ever appear together in two positions (you cannot have a time that reads 7:77, for example).

When I got to school, this faculty with numbers continued. During my early days in kindergarten, I sat at a desk by myself away from the rest of the class during the maths lesson. I was given eight- and ten-sided dice and was told to roll them and multiply the two numbers together. The first time I did this, when I showed the teacher the list of answers I had written down, she asked if I could go back and write down the two numbers I had multiplied together to get each answer so she could check my work. I thought her request was very weird.

My talent wasn't exactly nurtured whilst I was at school. I went to a standard primary school with no special classes or accelerated programs. But I reached second grade and still found myself somewhat ahead of the class. One time a substitute teacher was trying to teach us the concept of making an estimate. She wrote a series of fifteen different numbers on the whiteboard and asked the class to estimate how much we thought all the numbers added up to. The person who guessed the closest would win a prize. With a prize on the line, I added up the numbers in my head, which she had told us not to do. But, I thought, why estimate when you can work out the answer and win a prize? I remember 'guessing' (correctly) that the number was 158. Two other children in the classroom thought I might have made a mistake. They guessed 157 and 159.

By the time I went to high school, I was nothing special. I was decent at maths, but by no means was the teacher going to put me on a desk by myself. Where did that natural talent run off to? A few years back, I watched the movie The Man Who Knew Infinity and read the book it was based on. It is the story of the Indian mathematician Srinivasa Ramanujan. I am by no means comparing my talents to his as he was on another level altogether, but he had a natural talent for maths and it was hard for me not to feel a connection to that. His talent was nurtured at a young age. He spent most of his early days studying mathematics textbooks and solving problems on a chalkboard on his front porch. He went on to do amazing things in mathematics, and there is actually a discipline of maths devoted to deciphering his work.

Srinivasa Ramanujan's story started me wondering why his gift flourished and mine vanished. Why, by age thirteen, had I ended up with barely even a remnant of the natural talent that was so evident at age three? I asked my parents, and they couldn't recall that anything in particular had happened. They just said I lost it. As I began to reflect on my formative years, I started to agree more and more with their assessment – I had in fact lost my natural talent. But what concerned me was the thought that the talent may have been taken from me.

I grew up in what was the first generation of kids that could engross themselves in technology, and I consumed every piece of it I possibly could: television, Nintendo 64, Game Boy Color, Xbox, PlayStation 2, and computer games. Aside from television, all of these devices entered the market when I was a child. Nintendo 64 came out in 1996. Playing through every possible race in Mario Kart on every difficulty was a feat my brother and I quickly achieved. The Game Boy Color was launched in 1998. I recall spending hours playing through Pokémon Red and

Blue and trying repeatedly to finish Super Mario Land. The original Xbox came out in 2001, when I was nine years old. My brother got our Xbox 'chipped', which meant we could hire a game from a video store then load it onto our Xbox. For just the small cost of hiring out the game, we could play it forever. PlayStation 2 arrived in our house shortly after it was released in March 2002. Tekken 3, Spyro, Crash Bandicoot – you name it, we played it.

When it came to computer games, Age of Empires II, released in 1999, was a favourite with both me and my friends, and it would often take centre stage at LAN parties. For those who aren't familiar with the concept of a LAN party, where LAN stands for local area network, it was an event where participants would load their large desktop computer into the back of their parent's car, along with keyboard, mouse and every piece of equipment that connected to their PC and a few LAN cables, the cables you use now to plug into your internet modem, and transport the lot to the host's house. As I was frequently the host, it generally meant placing all the computers on my parents' dining table in a mess of cables and power boards, and then using LAN cables to connect computers together so we could play against each other. It was 'online' gaming in its most primitive form. The release of new games was anticipated with excitement. We'd plan events where we'd all gather at someone's house to watch the best gamer play through different stages of the game, and everyone searched the internet to uncover what tips, tricks and cheat codes had already been discovered.

But there was one game that was a part of my life for a very long time. So much so that my best man mentioned it during his toast at my wedding, having also brought it up at my eighteenth and twenty-first birthdays. It's a game called Guild Wars, released in 2005, that I signed up to in 2006. It's an MMORPG, which in non-nerd language means Massive Multiplayer Online Role-Playing Game. That means you get to play with people from all over the world and you do so by assuming the role of a hero. I still remember the early days of embarking on the journey with my chosen hero, doing missions and fighting bad guys along the way. It was a lot of fun, and I spent a lot of time on it.

There was a three-year period where I would have averaged a solid two hours every day playing Guild Wars. I know that because the game keeps track of the time you spend playing it. I've played more than 2600 hours at the time of writing, with over 1000 hours on just one character. 2600 hours is 110 days straight of playing. If I slept for eight hours a day and had only sixteen hours to play, that adds up to 162 days. That's essentially half an entire year of my life playing one game. It's a little hard to write, but it's reality. We will touch on the addictive qualities of technology and screens at a later point in the book, but clearly I had an issue with addiction and computer games. I recall many instances in my fourteen years of playing this game where I would quit in dramatic fashion, only to return to it later. I would delete characters, give away items and remove the game from my PC altogether. It was as if I had just quit taking drugs and was flushing them down the toilet. I was well and truly a glowface.

That was just one game though. I played countless other games along the way, and for countless hours. If I played Guild Wars for thousands of hours, I can't imagine exactly how many hours I must have spent in total playing games and watching television over the years. But it would be very safe to say, from a young child to my current age, I have spent over 10,000 hours watching television and playing computer games. Going off the eight hours of sleep a night assumption, that's 625 days of screen time – almost two years spent playing games and watching television. I use the number 10,000 because there's a theory known as the 10,000-hour theory. It states that you can be great at anything with 10,000 hours of work on that one thing – that it takes, at a bare minimum, 10,000 hours to truly master a skill. If you want to be a brilliant guitar player, it will take you 10,000 hours. Want to master a computer programming language? That will be 10,000 hours please. Think you're the next Picasso? It's going to cost at least 10,000 hours of your time to get anywhere near that. If I had wanted to foster my talent for mathematics and become a mathematician, it would have taken me 10,000 hours. But instead, I spent 10,000 hours killing bad guys, insulting newbs and watching television.

It's quite a hard reality to confront, the idea of what could have been. But it wasn't as if I was given a choice when I was younger to trade my future for present online gaming glory. No one confronted ten-year-old David and said, 'either you can keep this lifestyle of computer gaming and television watching, or you can utilise the talent you were born with and achieve something in mathematics.' I never consciously made that decision. I simply stumbled into each new device as it appeared and soaked up the radiance of its glow. It's understandable when you consider what putting a gaming device or a screen in front of a ten-year-old boy might do to his attention – that it might steal his attention away from everything else. But the reality of having spent a lot of my life staring at a screen manifests itself every day in my current career as a product manager.

It hasn't exactly been the most rewarding career. During my busy days of sending emails and attending Zoom calls, I can't help but ponder what I might have been had I spent my 10,000 hours in a different way. I have also begun to realise, that given I was a part of the first generation of children exposed to addictive gaming devices, I can't be the only one who regrets how they spent their

10,000 hours, and I can't be the only one who really didn't get a choice in how their 10,000 hours was spent. There must be others who, like me, feel as if some talent, some gift they had was taken from them in a way that they wouldn't necessarily have agreed to. That it was a trade they didn't know they were making.

It's quite a big claim to make, to say that excessive screen exposure leads to the erosion of your natural talents. The difficulty with the claim is in trying to measure scientifically to what extent screen time leads to the erosion of talent. How do you measure what might have been? How do you control for variation between individuals? But something that can be measured is how screens impact the development of the brains of young people. If there is a relationship between excessive screen exposure and some undesirable outcome in the brain development of young people, we can infer that in changing the structure of a child's brain, excessive screen exposure affects the development of their potential.

ABCD, it's as easy as...

The level of screen exposure in my generation pales in comparison with that of children today growing up with a smartphone in their face from the beginning of their life. Young people have never before been exposed to screens for such long periods of time at such an early stage in development. And although children have suffered from addiction to screens for decades, science is only just starting to explore the consequences of screen time exposure for children and adolescents.

The Adolescent Brain Cognitive Development (ABCD) Study is the largest study to date of its kind and marks one of the first real attempts at trying to better understand how the brain of an adolescent develops over time and how biological and environmental factors affect its development. The study is being undertaken in America and is funded by the National Institute of Health, with over US$300 million of funding provided to help ensure it achieves its goal. The study is tracking the cognitive development of over 11,000 children from the age of nine or ten until early adulthood. This involves taking scans of each child's brain at regular intervals throughout their development using MRI imaging to visually see the structural changes to their brain and how various environmental factors affect its development. The study is tracking many different factors, but one of the key goals is to examine the effects of screen time on brain development.

The initial set of scans of 4,500 children was completed in late 2018. When it comes to screen time and its effect on brain development, one of the key early findings by researchers was that there are differences in the structure of the brains in children who are exposed to screens for more than seven hours a day. Seven hours seems like a lot, but it can get racked up quickly. Add up all that time a child might be on their parent's phone, their own phone, an iPad or laptop at school or at home, watching television or playing video games, and seven hours starts to sound easily attainable.

The difference that researchers are seeing in the MRI images is that the cerebral cortex of the brains of children exposed to screens for more than seven hours a day are thinning much earlier than they should. The cerebral cortex is the outermost layer of nerve cell tissue of the brain, typically measuring only a few millimetres in thickness. It's responsible for thinking and processing information from the five senses. It is involved in memory, consciousness, motor function, planning and organisation, processing sensory information, language processing and determining both intelligence and personality. The cerebral cortex of the brain has a natural rate of thinning, typically beginning at the age of five or six. Researchers can't yet conclude that this accelerated thinning is solely based on screen time exposure, but there is at the very least a correlated relationship between screen time and the accelerated thinning of the cerebral cortex based on the findings from the ABCD scans.

Additional studies have shown that thinning in the cerebral cortex at a faster than expected rate has a relationship with decreased IQ scores over the same period of time. When researchers scanned the brains of children and compared the thickness of the cortex across a span of a couple of years, it was those whose IQ scores had decreased the most that showed the fastest rate of cerebral cortex thinning.

It's the job of scientists to turn hypothesis into theory, and theory into scientific law. So I'll admit it's a bit of a leap to infer that excessive screen exposure is making children less intelligent. It would be a big call to come out with the Law of Screen Time Dumbness. But it's somewhat difficult to ignore the possibility that exposure to screens is making children less intelligent.

Excessive screen exposure could also be impacting kids in other ways they are unaware of. The cerebral cortex doesn't just determine intelligence, amongst the many functions it is responsible for is personality. If excessive screen exposure is eroding young people's intelligence, it would be reasonable to assume it is impacting their personality as well. It could be affecting not just their ability to think, to reason and to remember, but who they feel they are as a person.

This is as close as we might ever come to showing quantitatively that excessive screen time erodes our gifts and talents. So to build this argument further let's consider some qualitative evidence and look at how those most in the know about the technology industry moderate their children's use of screens.

What the Outliers Say

Steve Jobs' children never used an iPad. Take that in just for a moment. The man who created Apple, the company whose devices most likely play some part in your life, didn't give his children the iPad. Bill Gates, founder and former CEO of Microsoft, took a similar stance, limiting his children's access to phones. He would not permit phones at the table when having a meal and didn't get them a phone until they were fourteen!

Mark Cuban, a high-profile American billionaire who amassed his fortune investing in media and technology companies, set similar limits on his thirteen-year-old daughter, making her turn her phone in at 10:00 pm on weeknights and 11:00 pm on weekends. He also admitted to giving his son money in order to not watch Minecraft videos. 'I paid my son $150 to not watch those videos for two months. But he could earn time if he watched maths videos, or did maths problems for me, he could earn time to watch Minecraft videos.'

Perhaps some of the most striking comments come from Alexis Ohanian, co-founder of internet site Reddit. Reddit is a social news website and discussion forum. Essentially, it can have information on anything and everything. It is full of noise and copious amounts of content that allow individuals to avoid boredom at all costs. So the comments from its co-founder regarding his plans to limit his daughter's screen time are striking. 'My wife and I both want her to be bored,' Ohanian tells CNBC. 'My wife and I both want her to know what it's like to have limits on tech ... I do look forward to playing video games with her when she's older, but it's really important that she gets time to just be with her thoughts and be with her blocks and be with her toys, so we'll be regulating it pretty heavily.' He wants her to be bored. He wants her to 'just be there with her thoughts'. These are very deliberate, very striking comments from someone who profits from a website that seems to provide people with the opportunity to do the complete opposite.

These high-profile individuals, all involved in the technology industry, have made very deliberate choices about how their children will interact with technology. They've set limits on technology that many of us would be afraid to attempt to impose on our children. They're motivated by a desire to have conversations with their children about books and history, to give their children the space to think and the opportunity for a good night's rest. For us to do some of these things might seem impossible. It might just seem too hard to operate in this world with a child who is fourteen and doesn't have a mobile phone. I tend to think though, if Bill and Melinda Gates, people who are constantly busy trying to change the world, can make the decision to impose those limits on their children, then we can set similar boundaries for the benefit of our children. Unless we set boundaries for our children, they will use devices designed to capture and hold their attention, and any boundaries they might set themselves will cease to exist.

I'm not saying you have to wait until your child is fourteen to give them a smartphone, or that they have to eat dinner at the table every night to be functioning humans. But it's clear that boundaries are needed. And these are examples of boundaries set by individuals who understand better than we do the danger that these devices present to young people's ability to develop their natural talents.

Parlez-vous français?

As a baby, my eldest daughter was very interested in, well, everything. In the first few months of her life, she resented being held with her face not looking out at the world. She had to be included in whatever was going on. This made keeping her entertained a difficulty. We decided to avoid any screen time for her until she was one, so we bought lots of toys. We bought way too many toys for our budget, but we bought them nonetheless to keep her entertained. What we saw, though, every time we went to a baby store or a store with baby toys, was a tablet device for kids. The device promoted educational applications to help kids learn and prepare them for school. I'm sure these are popular products because this is an idea constantly connected to technology – the concept that technology facilitates and enhances learning for children and infants. Often we justify the use of screens with children because the content they are engaging with is deemed 'educational'. In order to explore this, it is important to understand how children, and in particular infants, learn and remember.

My mother grew up in Cairo but is Italian by background. In Cairo she spoke five different languages – French, Italian, Greek, Arabic and English. When she came to Australia as a six-year-old, she mostly spoke only Italian and English, and the other languages were virtually left unused. In 2010, we took a trip to Europe together as a family, and our first stop was Paris. It was Mum's first chance to speak French in a number of years (I wouldn't dare say how many). Mum assured us she would be able to translate most things for us, which made the prospect of visiting a foreign country less daunting.

When we arrived, my mother tried to pick up what people were saying and respond in French. But we all quickly realised she wasn't going to fulfil her duties as family translator. Instead we ended up being the worst kind of tourists – not only could we not speak the local language, we had someone who thought they could speak the local language but couldn't.

After Paris, our next stop was Nice, a coastal city in the south of France. On the second day in Nice, we split up to explore some of the local shops and afterwards met up with my mother at a bakery. I'm not sure what happened, maybe it was a change in dialect, maybe her brain had heard the language enough that it fired up those dormant neurons, but when we walked into the bakery she was speaking fluent French to the lady serving her behind the counter. I was quite taken back, and I think she was too. After decades of speaking barely a word of French, suddenly she was conversing fluently. There was clearly something within her brain that lay dormant for years, and out of nowhere the switch was flicked on.

Researchers have spent a great deal of time trying to understand how the brain works, and they admit they have barely scratched the surface. There are just so many different fields of research – anatomy, physiology, cell biology, psychology, anthropology, to name a few – required to understand just this one piece of anatomy. The fact my mother suddenly remembered how to do something that she hadn't done in decades is hard to explain, but it could have something to do with how she first learnt French as a child. Children tend to be very good at what's called memory specificity. They remember things within the context in which they are learnt. If you show them a book time and time again, they know what is on the pages of the book and what the book is about. My in-laws love telling the story of how my wife, when very young, memorised the words of the book they used to read to her at bedtime. She couldn't read the words at that age, but her mind knew that, in this context, these words apply. Perhaps my mother's brain finally flicked the French switch because the bakery in Nice was the right context.

For children to transfer learning into a new environment, to generalise the details they have learnt into a new setting, they have to develop memory flexibility. Rachel Barr, a childhood development researcher at the University of Georgetown, says that '[memory] flexibility is crucial to the adaptability of learning and memory because it allows past-experience to be applied to a range of situations that are unlikely to be perceptually equivalent to the initial learning episode.'

For infants, their memory is very specific. It is because of this specificity that it is very difficult for young children to independently learn through two-dimensional materials such as information displayed on a screen. Let's consider an example. When a child is presented with a dog on a screen, the child is learning in that very specific context what it means for them to see a dog. If you keep presenting that two-dimensional image of a dog to the child, you will find that eventually they will understand that the object you are pointing to with black fur, long ears and a big tongue is a dog. But take that child outside and point at a dog walking past on the street and ask them what that is, and it's very likely that at first they will have no idea. I saw my daughter do this with colours all the time. She understood perfectly well which colour was which when we were drawing on paper using crayons and textas. But when I asked her to identify the colour of animals in a book, she would really struggle. Infants have difficulty in applying their memory to different scenarios.

Technically speaking, this inability to transfer knowledge is what is known as the transfer deficit. The transfer deficit describes children's struggle to learn when left to engage with two-dimensional media without any co-use. Two-dimensional media covers watching educational television programs or playing educational touch-screen games, but it also covers books. Placing books in the same category as screens may seem a bit counter intuitive. To adults, they clearly provide information in two very different ways. But to children, this distinction does not exist. Children simply greet both screens and books with the same level of engagement and curiosity, with pop-up books and books with flaps that hide pictures giving the interactivity of digital devices strong competition. Although books and screens both provide information to children in a two-dimensional format, the way in which we interact with our children when using books and screens to present information is often very different. As parents, we don't give a book to our young child and expect them to read it. We read the book to them, help them understand what is being seen and heard and then help them apply that in real life. Parents and children are using the book together; we are with our child, alongside them, helping them learn. The technical term for this is co-use.

Co-use of media is one of the ways in which we can help children overcome the transfer deficit of learning. Rachel Barr says that the three co-use techniques that can be used to improve infant memory flexibility are repetition and using visual and verbal cues. When we co-use media for educational purposes with children, we are inherently utilising those three tools. When we read a variety of stories about dogs to our kids and do it over and over again, showing them lots of different kinds of dogs in the stories and making the same silly sounding barking noise every time we see a dog on a page, we are helping their memories become more flexible.

And this is the problem with screens; they are seldom co-used for educational purposes. Typically, a screen is used as respite for the parent. Stick the child in front of a screen so that Mum and Dad can take a break for just a few minutes – a break Mum and Dad are more than entitled to. But don't mistake what it is. What you are doing is equivalent to placing a book in front of your child and expecting them to learn what a dog is from simply seeing a picture of one. No parent would ever expect their infant to learn from a book without help. But that's exactly the expectation we place on children when we let them engage with 'educational' screens on their own. Unless you are there guiding your child through the television show or helping them understand what is going on in the iPad game and telling them what it really means every time they get rewarded for progressing further, then the experience isn't educational. In fact, as we've seen in this chapter, such an experience for a child is most likely harmful to their intelligence, their personality and potentially their future life.

No More Product Managers

Before we end this chapter, it's important that I am transparent and honest with you. Both of my daughters didn't see any screens until after their first birthdays, but after that they have watched their fair share of television. We've tried to reserve the occasion mostly for when they were sick, but both weren't great sleepers and after a long night with little sleep sometimes they were placed in front of a screen playing Mickey Mouse Clubhouse. Both quickly understood that the phone was the device that made Mickey Mouse Clubhouse appear, and often after being exposed to it during a time of sickness they would both aggressively request it again for days after. One of my eldest daughters first television experiences was watching twenty minutes of *Finding Nemo* and for several days afterward she would make the sound 'blub blub', move her hand like a fish and point towards the television, indicating that she wanted to watch *Finding Nemo*. What struck me was how quickly she created a level of attachment to the screen. As she moved into being a toddler it was hard to kick the addiction, but my wife and I have gone a long way to setting boundaries on television and ensuring that as a family we engage with it in an appropriate way.

I think this is a small example of the much larger challenge I will face as a parent moving forward. My daughters will be exposed to more technology and more screens as they grow up. I know I cannot prevent their engagement with these devices entirely, and I know trying to do so would be bad both for their development and for our relationships. But what I do know is that I can help them to understand the potential danger that these screens bring to their lives. I know I can encourage development of whatever talent arises in them at a young age, providing them with every opportunity I can to help them excel. Ultimately, my goal is to ensure they do not trade their innate talents for screen time because that's not a trade they can decide to make for themselves.

Children are unaware of the trade that is being made every time they are placed in front of a screen. From only a few weeks old, children are handed phones to play with and screens to touch. Infants are exposed to iPads, to games on their parents' phones, to Netflix – to anything that will entertain them. We as parents, family members, caregivers and educators, give into this for many reasons. But what we need to realise is that we are making a trade on behalf of these kids. These screens are literally taking gifts, talents, dreams and intellect from our children, and it isn't fair that they do not get the chance to decide whether to make that trade. The reality is we might be trading away our child's chance to be a doctor, a lawyer, a scientist, or a mathematician, and they might just end up working as a product manager.

Chapter 2

The Loss of Depth

We will make the whole universe a noise in the end.
C.S. Lewis

I've spent many years travelling into Sydney's central business district for work. Depending on where I lived at the time, the train journey would take anywhere between thirty and fifty minutes. After regularly commuting to work, I began to understand that there are a few unspoken rules of travelling on the train in Sydney. One rule I live by every day is to make a very conscious decision about who to sit next to. I wouldn't normally encourage people to 'judge a book by its cover', but on Sydney trains it is a necessary survival tool.

I tend to apply a few criteria to my decision. Firstly, the person cannot be eating food. I will never understand how someone has time to put their breakfast cereals into a container, pour the milk in and close the lid, but not make time to eat it whilst in the confines of their own home, instead choosing to ingest their cornflakes during a bumpy, rocky train ride. My worry about sitting next to them is how bad I would smell for the entire day if that person managed to spill cereal and milk on me – regardless of the kind of milk this week's diet calls for. I also don't like people who I know I'll be fighting elbows with. The train is a means of getting to work, not a place to set up your office, conduct conference calls and smash out emails with your elbows jamming into the side of whoever makes the mistake of sitting next to you. My final standard is a bit harder to judge, but it is very important. If I think you are going to smell, there is no way I am sitting even remotely close to you. I feel it is common courtesy for the general public that individuals at the very least apply deodorant and wear a clean set of clothes each day when travelling on public transport.

One type of person it is almost impossible to avoid is the person with a screen, particularly a phone. The majority of individuals travelling on the train to and from work spend their time on their phone. Whether they are aggressively scrolling and tapping their way through Instagram or are binge watching the latest show released on Netflix, the phone occupies the space in front of their face for the majority of their work commute. Perhaps these individuals have a long stressful day of back-to-back meetings and non-stop interruption and want a diversion beforehand, or they have just completed such a day and are simply looking to unwind. Whatever the reason, the phone is the most popular means of passing the time.

Getting sucked into the vortex of my phone is one trap I try to avoid at all costs. As I have spent most of my day staring at a screen at work, the notion of doing more of it isn't appealing. My train trips have always been long enough to get something productive done, but just short enough for me to kill the time doing nothing if I wanted to. Reading is my first option, but if it's been a long day at work and my brain feels pretty fried I try to listen to music or a podcast to unwind. But to say I've never fallen into the phone vortex would be a complete lie. We all get to the point in the day where we've had enough. We are done thinking and we just want to unwind. We are now entitled to 'turn our brains off', to 'veg out'. We have used our brain throughout the day to problem solve, compose emails and navigate politically charged meetings. In corporate environments filled with constant distraction, the way in which our brain is used is exhausting. But it isn't exhausting in the way it should be.

Missing My Brain

I graduated from university in 2014, completing a Bachelor of Commerce degree. I'd spent four years at university. After starting university wanting to be a doctor but failing to get the grades, I swapped over to a commerce degree as it was a way to get into banking. My dad had spent many years working for a bank and my brother also worked for a bank, so I figured I could find a fulfilling career and earn lots of money doing the same.

I enjoyed what I studied at university and during the degree heard people talking about getting graduate roles with large investment banks. I tried to follow a similar path and applied for every internship I possibly could. I didn't get a single interview. The people who tell you your marks don't matter are the people who are testing you to see if you believe them. Unfortunately, I believed them.

Once university finished, I found myself starting out in an entry level role at a bank, assessing mortgage applications. In this role I essentially worked in thirty-minute intervals. During each interval I would determine if the applicant could borrow what they were asking to buy their property. I'd look at how much they earned, see how much debt they had and then make a decision about how much they could borrow based on a balance of those things. Once I'd made a decision, I'd move on to the next application and do it all again. I started out bright eyed and optimistic that this would lead to a fulfilling career. Unfortunately, it didn't.

What I often found during this time when I was fresh out of university, was my colleagues and I would often discuss our feelings about having finally reached the point in our lives when we were no longer students. Initially I was relieved to have left behind lecturers I couldn't understand, deadlines I felt I couldn't meet and a life without weekends. Getting paid and having money felt great – for about a month. Then I began to become more and more disenchanted with the daily routine, and I found myself going through the motions. I very quickly discovered I missed university, missed the difficulty presented by lecturers I couldn't understand and deadlines I felt I couldn't meet. Fundamentally, I missed the opportunity to think, and I started saying this to my friends and colleagues when discussing my university experience. I missed university, I would say, for the very reason that it was difficult, for the very reason that it forced me to use my brain.

A corporate life is a life spent heavily in front of screens. In corporate roles you are constantly inundated with information on the screens in front of you. My first role in the corporate world was no exception, and I began to wonder what this was doing to my brain and why I felt as if I wasn't really using it. Writer Nicholas Carr went on a similar journey, which he details in his book

The Shallows. He experienced the rise of technology in both his profession and his personal life, but found himself at a point where he was missing something he felt was crucial:

> *Sometime in 2007, a serpent of doubt slithered into my info-paradise. I began to notice that the Net was exerting a much stronger and broader influence over me than my old stand-alone PC ever had. It wasn't just that I was spending so much time staring into a computer screen. It wasn't just that so many of my habits and routines were changing as I became more accustomed to and dependent on the sites and services of the Net. The very way my brain worked seemed to be changing. It was then that I began worrying about my inability to pay attention to one thing for more than a couple of minutes. At first I'd figured that the problem was a symptom of middle-age mind rot. But my brain, I realised, wasn't just drifting. It was hungry. It was demanding to be fed the way the Net fed it – and the more it was fed, the hungrier it became. Even when I was away from my computer, I yearned to check email, click links, do some Googling. I wanted to be connected. Just as Microsoft Word had turned me into a flesh-and-blood word processor, the Internet, I sensed, was turning me into something like a high-speed data-processing machine ... I missed my old brain.*

The brain I had at university was allowed to think and solve problems. The brain I now had in a corporate world was reacting to whatever the screen in front of me told me. I too missed my old brain.

What Screens Are Doing to Our Brains

If excessive exposure to screens is impacting the lives of young people, including changing the trajectory of their journey through life, then the potential for excessive screen exposure in adulthood to have equally dire consequences needs to be explored. We saw that the negative effects of excessive screen time was most evident in children exposed to seven hours or more of screen time per day. For any of us who work a corporate job, this amount of screen time is easily reached on a daily basis. Philosopher Frédéric Gros brilliantly describes the life of a corporate employee as they make their way through a standard workday:

I think of those abstracted sedentary individuals who spend their lives in an office rattling their fingers on a keyboard: 'connected, as they say, but to what? To information mutating between one second and the next, floods of images and numbers, pictures and graphs. And after work it's the subway, the train, always speed, the gaze now glued to the telephone screen, more touches and strokes and messages scrolling past, images ... and night falls, when they still haven't seen anything of the day. Television, another screen.

Even for those who don't spend their day in an office, accumulating a large number of hours looking at a screen each day comes easily. Most people regularly look at their phone during the workday, whether between tasks or during meal breaks. Once home, people often unwind by watching a few hours of television, scrolling through social media or exploring an alternate world in a computer game. Racking up seven hours of screen time each day is easy when you consider our access to screens and the prominent place they have in all our lives. We are surrounded by screens regardless of our industry, and our interaction with them can leave us missing our old brain – the brain we had before it was moulded into something different through a function called neuroplasticity. Before we define neuroplasticity, we must first explore some basic parts of our nervous system.

Our nervous system works by sending electrical signals across cells called neurons. Neurons are the cells within the nervous system that transmit information to other neurons or to other types of cells in our body, such as muscle cells. When our brain sends a signal to a body part to perform a particular action, such as kicking a ball, it does so by sending an electrical signal along multiple neurons to reach the muscle cell and activate the cells in the desired fashion, in this instance to kick the ball. These multiple neurons form what is known as a neural pathway.

When any neural pathway is first activated, the signal is very weak and the effectiveness of that signal is poor – our body doesn't do exactly what our brain wants it to. However, if that neural pathway is utilised over and over again, then the neurons become more effective at passing that electrical signal and our body gets better at doing what our brain wants it to.

An example I often draw on is how a person's body reacts when they first go to the gym. When someone attempts a bench press for the first time, as they start to take the bar off the rack and move it towards their chest, what often happens is the muscles they are recruiting appear to tremble. It's almost as if the muscles are in spasm. Although the weight on the bar may be light, first timers will often have trouble controlling the bar as they lift it and lower it to their chest. This is typical of any standard gym movement or exercise, such as a squat or an overhead press. The first time someone performs this exercise, their muscles will invariably move uncontrollably in some way. What's occurring here is the nervous system is reacting to a stimulus it has never encountered before. You may mentally understand what's required in a bench press, moving the bar off the rack and touching your chest with it before moving the bar back up to where your arms are locked out, but your nervous system isn't exactly sure which muscle cells to activate at which time. So your nervous system goes a bit crazy and fires lots and lots of signals in the hope of getting the general movement correct. But those millions of signals result in a movement that is a bit shaky or wobbly because the muscle cells are firing when they don't need to and in an order that isn't correct. The nervous system is trying to learn how to perform the movement. It's trying to form a new neural pathway.

Neuroplasticity refers to the ability of the nervous system, in particular the brain, to re-purpose these neural pathways if needed. It is a relatively new idea within neuroscience. Up until the 1970s it was thought that the brain was fixed when we reached adulthood. That is, however the brain looked in our mid-twenties and early thirties was how it was going to look for the rest of our lives. But what in fact happens within the brain is, if we continue to use a neural pathway, that pathway will get more and more entrenched. It will become stronger and the signals will flow faster. Think of a stream of water that starts flowing along the ground. At first it will take a variety of different paths and the flow won't be very strong. However, as the water continues to flow, it will wear away at the ground and create a channel, and the speed and strength of the flow of water will increase. So, too, with our neural pathways. This is exactly what's happening in the bench press scenario. As the signals all start moving along the one neural pathway, the speed and strength of the signal increases and so the movement of the bar becomes more controlled.

This whole idea feels somewhat intuitive. We've all heard the expression 'practice makes perfect' – the more we do something the better we get at it. This is neuroplasticity at its most basic level and what it will hopefully make you appreciate is that the physical structure of our brain is changing constantly. Neural pathways are being strengthened or re-purposed all the time. This isn't just happening in things we specifically choose to do in very deliberate ways, such as learning a new language, playing an instrument or perfecting a bench press. Neuroplasticity is a process that is happening inside our brains all the time, including every time you look at a screen.

The temptation is to think that it takes a long time for our brains to change, for new neural pathways to form and to strengthen and for old neural pathways to die off. You could think that given it takes a long time to learn a new skill that it would take a similar amount of time to rewire your brain. But change happens fast. In fact, research has shown that screen use is a strong catalyst for change brought about by neuroplasticity. One experiment showed that it took just seven days for the brain to be re-wired in response to just one hour a day of internet use. This experiment is a great example of our brains being very 'plastic', of them being easily moulded and changed based on external stimulus. Neuroplasticity is a constant process, and technology has advanced such that we now have constant access to stimuli, like the internet, that can rewire our brains. When you combine the continuous nature of neuroplasticity with the pervasiveness of screens, you begin to see that screens have an unprecedented ability to reshape how we think and make us think the way they want us to.

Screens want us to think in short sharp bursts of attention. All of the screens in our lives are designed for the primary purpose of capturing and keeping our attention. The majority of applications on each of the devices we use are designed to pull us in, to get us to look at them and to keep us engaged. This is because attention has become the world's most valuable commodity. The longer companies can keep our attention, the more profit companies can make from the content displayed on the screen. Consider one of the most popular social media platforms ever, Instagram. As a user you don't pay to use the product. This is an unusual idea. Most products and services require payment, that's how companies generate profit. But social media does not require a monetary investment from you. You don't pay to be able to connect to friends and share photos. Instagram freely provides one of the most well-designed digital products ever created. What is of value to Instagram is understanding each user and displaying relevant marketing content to them in the hope that they engage with it and make a purchase. The more time a user spends on Instagram, the better Instagram understands the user and what they like and dislike, so the more tailored Instagram's advertising can be to them and the more profit they can make from advertisers because they know with greater certainty if the content shown will result in a sale. So, not only is Instagram itself a distraction with friends' photos and updates, but the user is also being bombarded with attention-grabbing advertising material in an attempt to keep them constantly engaged. This is just one example of the many digital applications that constantly fight for our attention.

Screens are designed to distract you. Even the iPhone tracks how many times you pick up your phone each day in an attempt to let you know that something is wrong. As you'll come to see throughout this chapter and the book, this constant fight for attention rewires the brain to thinking only in short bursts of attention. Thinking only in terms of headlines, quick replies to text messages and clicking as fast as we can on notifications. Our brains are being moulded to be very good at being distracted, at shifting our attention between activities every few seconds. We therefore become very bad at the opposite activity, giving our attention to a single task and focusing on it for a long period of time.

This is one of the most significant changes that has resulted from the combination of neuroplasticity and constant screen usage. It turns out that not only are we being forced to swap our attention between tasks by screens, but we also have a natural disposition for this very activity, making it even easier for screens to shape how we think.

The Distracted Brain

Daniel Kahneman is a Nobel Prize winning psychologist. Among his many achievements, Kahneman, along with fellow psychologist Amos Tversky, introduced the world to the idea that individuals often make decisions that don't fundamentally make sense. Kahneman and Tversky showed that humans are susceptible to decision-making biases such as representativeness, availability and anchoring. We'll look at just the first of these, representativeness, using the Linda Problem.

Consider the following description of a woman called Linda: Linda is thirty-one years old, single, outspoken, and very bright. She majored in philosophy. As a student, she was deeply concerned with issues of discrimination and social justice and also participated in antinuclear demonstrations. Which alternative is more probable?

1. Linda is a bank teller?
2. Linda is a bank teller and is active in the feminist movement?

Take a moment to answer the question before moving on.

Kahneman reports that between eighty-five and ninety per cent of undergraduate students at several major universities across America chose the second option. But if you think in terms of a Venn diagram you understand that the second option is far less probable than the first.

The set of feminist bank tellers is included completely in the set of bank tellers – in order to be a feminist bank teller, you must first be a bank teller. Therefore the probability that Linda is a bank teller must be greater than the probability that Linda is a bank teller AND a feminist.

Chances are you chose the second option. This is because of the way the description of the woman is presented, hence the term representativeness. The problem depicts a woman who might typically be considered a feminist. So, upon reading the description of Linda, your brain probably concluded she was a feminist before you were even presented with the two options. Your brain wasn't really concerned with the question itself, which asked 'which alternative is more probable?', it just concluded that Linda was a feminist. Before you knew it, you were like the ninety per cent of undergraduate students who fell victim to the decision-making bias of representativeness and misjudged the probability of the situation.

Kahneman would say that you've made this mistake because you used your System 1 rather than your System 2. In his book, *Thinking, Fast and Slow*, Kahneman tells the story of the ongoing struggle of the two characters of our brain, which he calls System 1 and System 2. He says that our 'System 1 operates automatically and quickly, with little or no effort and no sense of voluntary control' whereas our System 2 involves 'allocating attention to the effortful mental activities that demand it, including complex computations. The operations of System 2 are often associated with the subjective experience of agency, choice, and concentration.' He argues that when System 2 is engaged, human beings are most attentive, most alert to their surroundings and least prone to succumbing to decision-making biases. It is therefore our System 1 that is responsible for our biased decision-making. We tend to engage our System 1 because it takes the least amount of effort. In comparison, using System 2 is hard work. Kahneman is very deliberate in his description of System 2, saying it requires effort, is used during complex computations and involves a deliberate allocation of our concentration. In fact, Kahneman argues that using our System 2 burns calories, and as we deplete our calories, we tend to want to use our System 2 less.

Kahneman refers to a study of eight parole judges in Israel. The parole judges spent their entire day reviewing parole applications. They took an average of only six minutes to review the merits of each case and determine whether or not to approve parole. Parole was approved on average only thirty-five per cent of the time by these judges. In the study, the researchers recorded the exact time of day each decision was made and also noted when the judges took breaks for food. What the researchers found was that right after a meal, the approval ratings soared to an average of sixty-five per cent and steadily declined until the next meal. The approvals declined so much that the approval rating of the parole case considered right before each judge would take a break for a meal was essentially zero. If you were the unlucky individual to be considered before mealtime it didn't seem to matter whether your parole application was of merit or not. The judge's ability and desire to think had been diminished to the point where the effort to do so had been abandoned and the easy decision to deny parole was taken. By the end of the approval session, the judges had given up using their System 2 and reverted to the 'non-thinking' System 1.

System 2 requires effort, unlike System 1, which does not. We typically use our System 1 because it is inherently the easier option. As Kahneman puts it, 'Laziness is built deep into our nature'. Our desire to conserve energy, to apply the least amount of effort, to think as little as possible when making a decision, is built deep into our nature. And the screens in our lives feed off this natural preference.

When we engage with our screens, we are most often using our System 1. We are not really thinking actively about the content we are consuming. We are not deep in thought about whether we should like a photo or reply to a tweet. We are not considering each time we lift our thumb whether we should repeat the action and scroll to view the next piece of content. Even in Words with Friends, a game that appears to require thinking, we often just throw random letters out onto the screen in the hope that we magically stumble across a word we have never heard of. It is this innate desire to use our System 1 that can increase the negative impact resulting from the rewiring of our brains from screen usage. We are constantly inducing our brain to use System 1, constantly encouraging our brain to take the easier path and not think. Our brains are essentially okay with this because thinking is hard. But thinking, as you will see, is worth it. It is arguably the most valuable, the most human thing you can do. Thinking, using our System 2, is where we flourish as humans. But before we get to that, there is another issue we face when dealing with screens, and that is their addictive pull.

The Distraction Addiction

Chances are you've heard of dopamine. It's a neurotransmitter used by the brain to reward an individual for fulfilling a need or desire. When the brain releases dopamine, it makes us feel good. This makes dopamine one of the most critical elements in the addiction process – if something triggers the release of dopamine in our brain, making us feel good, then chances are we will want to repeat that process again and again. A very basic example of an activity that releases dopamine in the brain is eating. Eating is a necessary part of our survival, so our brains encourage this behaviour by releasing dopamine whenever we eat. Certain foods have properties that promote increased dopamine release.

It will probably come as no surprise to you that one of those foods is chocolate. Chocolate contains small amounts of a compound called phenylethylamine. This chemical stimulates the release of dopamine within the brain. Chocolate also contains a compound called tyramine, which has a similar dopamine increasing effect. These two compounds are part of the reason why it is so easy for us to overindulge in chocolate. It is also why we often find ourselves craving it almost instantly after finishing a meal. Fortunately, we have a feedback mechanism that inhibits us from continuously getting dopamine releases from chocolate – our stomach. When our stomach is full, we can't get any more dopamine from eating chocolate, so we take a break. In fact, with any natural human activity that releases dopamine within our brains (sex being another example), there is generally a feedback mechanism that stops us from continuously getting a feel-good dopamine hit.

Eliciting dopamine from the user is a basic premise of any good screen based application. One of the most prominent categories of dopamine inducing technologies are video games. If you are familiar with video games, you will understand how they are all designed around this process of rewarding the user in the hope of eliciting dopamine. Games often reward users for logging on, and they are built around easily obtainable achievements and continuous progress to unlock new and fun things for users and keep them engaged. Every time the user unlocks something new, each time they reach a new level, there is a dopamine hit waiting for them. The brain quickly learns that this hit is easy to obtain, so it keeps playing the game to get more dopamine. I haven't ever actually played Candy Crush, but I've caught myself many times on the train staring at someone's phone as they play the game. I've got no idea how it works, but I can see that with every swipe of the thumb another hit of dopamine awaits as the screen lights up, tiles explode and points are earned. Game designers aren't stupid, this is how they keep people playing. At the height of my addiction to video games, I'd wake up in the morning and that's all I could think about. The sensation I'd get when I finally logged on to the game and started my adventures was incredible, and throughout my time on the game I'd feel such a sense of reward when I achieved things or progressed further. My addiction was real, and I see now how dopamine played a large part in my overuse of those games.

The Adolescent Brain Cognitive Development (ABCD) Study is also looking at the relationship between screen addiction and dopamine. A 2018 report by *60 Minutes* detailed some of the findings of the study. In particular it referenced the results of an MRI scan performed on a young participant named Roxy. During the interview she acknowledged that she would check her phone every ten to twenty minutes during the day, admitting it was probably a conservative estimate. When performing the MRI on Roxy's brain, they placed her phone in another room but set up mirrors so that she could view it. They then put up her Instagram feed and started to scroll through, all the while still having her brain scanned by the MRI. Immediately, the researchers saw activity in her brain's pleasure centre, indicating that dopamine was being released into Roxy's system as she viewed her social media feed. Roxy didn't even need to be touching her phone for it to give her pleasure.

The control these devices have over our mind is often overlooked. But examples like this force us out of the denial that they aren't a problem and into the reality that even seeing them from a distance can elicit a response from our brains that reinforces our addiction.

Video games and social media, often the two biggest components of a person's screen time, have been shown to stimulate the release of dopamine in our brains. What these two also have in common is that interactions are fundamentally built around novelty, another driver of the release of dopamine. The novelty of the video game or social media experience is the excitement that comes from the ever-present opportunity to find something unusual, something new or something that alleviates boredom.

Winifred Gallagher, author of *New: Understanding Our Need for Novelty and Change,* argues that 'our genius for responding to the new and different distinguishes us from all other creatures'. Gallagher notes that from the time we are babies crawling around, we seek out that which is new and different. When my first daughter started crawling, it was all about exploration. Where can I go that I haven't been yet? What can I touch that I haven't touched yet? What can I put in my mouth that I haven't tasted yet? We tried to set her up in her room with toys and keep her in there by closing a baby gate, but her interest in the toys would last only a few moments, then she would be at the baby gate demanding the freedom to explore.

We are hard-wired to respond to things that are new and different in our environment and to have a curious, inquisitive nature. It's our ability to respond to that which is new and different, to react to the unsuspecting sound or the unfamiliar sight that has helped us survive as a species. From the standpoint of evolution, this makes perfect sense. The creature that is most attuned to its surroundings, most ready to react to change in the environment, is the one most likely to be able to quickly notice danger and react accordingly. The creature that is able to do this whilst conserving the most amount of energy, whilst not using calories, is the one best positioned for survival. As Nicholas Carr explains:

The natural state of the human brain, like that of the brains of most of our relatives in the animal kingdom, is one of distractedness. Our predisposition is to shift our gaze, and hence our attention, from one object to another, to be aware of what's going on around us as much as possible.

It's this survival mechanism that is being taken advantage of when we use our screens. Some people return constantly to social media because they receive a dopamine hit from the likes and the social approval. But lots of people use social media without posting any content. They are on social media for the novelty: the cool story their friend posted, the interesting news article their relative shared. Whenever we experience new and novel things on social media, we are being conditioned to return to that platform to search again for novelty. When you combine novelty with dopamine you get a powerful mix of two very strong factors of addiction that make putting your device down very hard.

It's no wonder that screen addiction has become a real issue in our society. Internet addiction disorder is a clinically diagnosable problem that needs professional help and rehabilitation facilities dedicated to people struggling with technology addiction are very much a reality of our present society. Stories of video game addiction fill Dr Kardaras' book *Glow Kids* – stories of families whose lives have been ruined by children obsessively playing video games. Dr Kardaras recalls his interactions with a young boy named Mark, who's mum Cathy, had emailed him asking for help:

Mark, who had started fiddling with a computer when he was five because his well-intentioned mother thought it could be educational, fell into a horrible screen addiction that was destroying his life. From the time he was very young, his mother told me, his whole demeanour would shift when he got in front of a screen ... Once he discovered video games at the age of ten, it was all over. He would steal from his mother to buy video games and game consoles; he would get violent and aggressive when he wasn't allowed to play; he lost all interest in school and hobbies that he had loved.

Dr Kardaras says that Cathy defined her son's interaction with screens as 'just like any other addiction'. It's a dramatic word, addiction. But addiction is a very real risk with screens. An addiction is something you cannot control – something so out of control it has an adverse impact on your ability to perform the responsibilities you have to yourself and to those around you. In Mark's case this definition clearly applied. If you apply that definition to someone constantly on their phone or a screen and compare that to someone who is addicted to drugs or gambling or any other substance or activity, you reach a very similar outcome. Someone who is addicted to illicit drugs cannot do their job properly. They cannot be a husband or wife, or a father or mother in a way that is sufficient for the responsibilities demanded of them in those roles. They cannot go more than a few moments without reaching for drugs to give them a high. They are controlled by the drugs. They are controlled because they have a physiological craving for the substance they are taking. With screens, that substance is dopamine.

There is a good chance you've felt this way with your phone or a screen before – controlled by it. This is because your phone or a screen can trigger the release of dopamine in the same way you can get a stimulant hit from drugs. You probably can think of a few people in your life who you know are controlled by screens, who are addicted to screens, who let screens get in the way of their responsibilities to their colleagues and to their families. But people frequently don't take this seriously. It is my hope in this chapter and throughout this book that we might raise the awareness of screen addiction as a real issue impacting the lives of millions of people each and every day.

Unable to Think

Whether it's their appeal to our natural inclination to the easy path or through blind addiction, screens are changing how our brains work. They are making our brains operate in simple, reactive ways, and what is ultimately being lost is thinking – deep thinking that results in a proper understanding of whatever it is we are considering. We cannot truly think whilst we are living in a world where we are interrupted every five minutes by a notification or email. We cannot truly think when we are constantly inundated with the voices and opinions of others. How can we really form our own opinions, define our own values and discover our purpose when we never provide ourselves with the opportunity to really think those things through? This kind of deep, concentrated thinking requires being engaged with the activity at hand. You're present, mindful and active in your consideration of the topic. It's when your brain has completely turned on its System 2 and is ready to make an informed, carefully thought-through decision.

The author Cal Newport examines this kind of thinking, which he calls 'deep work', in a book he has named after the concept. He defines deep work as 'professional activities performed in a state of distraction-free concentration that push your cognitive capabilities to their limit'. These efforts, he says, 'create new value, improve your skill, and are hard to replicate'. To be in a state of deep work is, for Newport, the best place to be when looking to produce one's most fantastic work. Key to Newport's definition is the creation of 'new value'. It's in this new value that the individual finds the most fulfilment and satisfaction from thinking. Keep that in mind – when you are doing deep work, you are most fulfilled and satisfied.

The psychologist Mihaly Csikszentmihalyi also offers a view of thinking that prizes deep engagement, a concept he calls being in a state of flow. Flow is described as 'the state in which people are so involved in an activity that nothing else seems to matter.' Csikszentminhalyi see's flow as the most 'optimal experience' a human can have. It occurs when individuals find the optimal balance between the difficulty of a task and their level of skill. Too difficult, and the task will be too demanding and the individual will give up. Too easy, and the task will be too boring. It's essentially where an individual is 'in the zone'. Being in the zone is where you are most attentive to the present task at hand. It's where outside distractions do not phase you and you don't go looking for any distractions for yourself. Being in a state of flow is a powerful way of describing the times in our lives where we are truly thinking. Csikszentmihalyi comes to a similar conclusion to Newport, arguing that individuals find optimal experience when they control the content of their consciousness, when they are able to focus on what they choose to focus on. They find optimal experience when they are not distracted, when they are in a state of flow.

The screens in our lives are taking away that which can bring us our truest joy – thinking. Bill Gates was asked in an interview what his biggest fear was. He responded by saying that his biggest fear was losing his ability to think. Thinking is valuable, more valuable than all the money you could ever dream of. Learning to think independently, flexibly and creatively are traits in life that will enable an individual to live a life of purpose and to foster a meaningful existence. To be able to encounter situations that are difficult, that require System 2 thinking, and not shy away from them. Rather, to overcome them by engaging with them, concentrating on them and working out a solution.

Too often we encounter problems and shy away from them or seek distraction to avoid dealing with them. How often have you found yourself faced with a difficult decision only to instinctively reach for your phone? That's your mind quitting before it really even had a chance. It's seeking distraction, seeking the easiest way to turn on your System 1 and get a hit of dopamine along the way. A life best lived is a life lived embracing those difficulties, doing deep work, and finding a sense of flow – being deliberate in our actions and being mindful in our decision-making. Too often we use screens in ways that are not deliberate, in ways that are mindless, because that is what screens reward. I encourage you to take back control of your mind, to put away the screen and start to think again. If your brain has been changed by constant hits of dopamine, it is also capable of changing back. That is the miracle of neuroplasticity. Don't let yourself think you're too far gone, that the changes to your brain are fixed in concrete. It's ready to be changed back to the brain you had before screens took control. It will be a journey, but when you find your old brain, you'll never want to let it go again.

Chapter 3

The Loss of Careers

Connection is inevitable. Distraction is a choice.
Alex Soojung-Kim Pang

At the time of writing this book, I worked for a company that went to great lengths to retain their employees and keep them engaged. Along with providing additional benefits such as subsidised healthcare and gym memberships, the business also holds many events internally that make it a unique place to work. One such opportunity that was presented to me during my first year was a Zoom call with the CEO of the Australian business that took the form of a Q&A session with him and a small group of other employees.

I had a lot of respect for this individual. Unlike some other heads of business that I've worked for, he was articulate, thoughtful and honest. He held regular company wide meetings called town halls, and he seemed adept at answering a range of questions. The prospects for an interesting Zoom session seemed high. Town hall meetings in previous businesses where I had worked were capped by everyone shouting 'onwards and upwards' at the command of the head of the business. I could rest assured my small Q&A session wouldn't end with such a meaningless phrase.

The Zoom session was an opportunity to advance my cause, the fight against distraction, with the CEO. So, I blocked out time in my calendar to plan my question and ensured I had all the necessary statistics ready to back up my claims. I made sure I wore a nice shirt, shaved and did my hair – which is quite the trifecta for a Zoom call. The call started and the CEO took questions from the group straight away. I wanted to make sure my preparation wasn't wasted, so I jumped right in asking him why the business doesn't invest time and money in educating staff on the dangers of distraction in the workplace. I tried to appeal to commonalities between us all, such as the constant interruptions from our phones and our kids. But, to be honest, I was too nervous and rambled on a bit too long for him to really answer the question. In effect, his response was that everyone is an adult and should be responsible for policing their own time. It was a response to be expected from someone who himself must have been a very diligent worker to achieve the success he has. A few others in the group asked if I had used the *Headspace* meditation app. Too ashamed to admit I used that app almost every day, I abandoned ship, thanked them for their insight and put a stop to the question. Unfortunately, no additional training plans were made as a result of my well thought out but poorly executed question.

Companies spend millions of dollars on employee training. They cover everything possible, from role specific training to providing opportunities for those who wish to expand their horizons. But few, if any, companies appear to be aware of the impact of distraction on their employees, and how educating them on the dangers of distraction and helping them to manage these distractions would greatly benefit their businesses.

Distractions in the Workplace

The online education company Udemy conducted a survey of 1000 professionals and published the results in the *2018 Workplace Distraction Report*. The survey sought to 'measure how distracted employees are during work hours, how they're responding to distractions, and what it all means for employers.' The report separates out the results based on the generation of the responders. This breakdown gives us insights into what one generation might be doing better than another when it comes to dealing with distraction.

Udemy found, for example, that distraction seems to affect the younger generations more than the older ones, with seventy-four per cent of millennials (individuals born between 1981 and 1996) and generation Z (individuals born between 1996 and 2012) respondents reporting that distraction was an issue in the workplace. We might think that these generations are the ones best equipped to deal with technology. They've grown up with it, after all, and had the most time to understand how it might influence their day-to-day lives; surely, they are the best equipped to handle technology in the workplace. The evidence tells a different story. It's those who have never known anything else besides technology who know the least about how to deal with it. This reality is highlighted throughout the report, with thirty-six per cent of millennial and generation Z respondents saying they spend over two hours a day on their phones whilst at work and sixty-two per cent of them saying they spend at least an hour a day on their phones whilst at their jobs. That's two out of every three young employees spending almost one full workday each week on their phones.

The biggest culprit for distraction appears to be personal activities, with seventy-eight per cent of these young respondents saying they are more distracting than work-related interruptions. Of these personal activities, social media is noted as the biggest problem, with either Facebook, Instagram, Twitter or Snapchat being labelled as the number one online distraction by eighty-eight per cent of millennial and generation Z respondents. What's the result of all this distraction? Forty-six per cent say that distraction in the workplace makes them feel unmotivated, and forty-one per cent say it stresses them out. Distraction is making young people within the workplace unhappy.

But the report shows that it's not just young people. Fifty-four per cent of all respondents say that distractions are affecting their performance, and fifty per cent say they're significantly less productive. When distractions are reduced in the workplace, three out of four employees say they get more done and forty-nine per cent say they are happier.

So, essentially half of all employees say they are happier at work when they are less distracted. It's a result that is in line with what we've learnt about the value of distraction-free work – being in a state of 'flow' or doing deep work. We know that these kinds of activities result in focused concentration and ultimately produce the most amount of satisfaction. Without even being prompted towards these styles of work, half of all individuals working in a corporate culture have a preference for it. There is a deep, embedded desire within our brains for focus, and the distractions that bombard corporate professionals prevents them from truly focusing and delivering on their work.

Where Does Work Get Done?

There have been attempts to improve the quality of work in corporate culture in recent years through the introduction of digital 'productivity tools' within the workplace. These include applications such as Slack, Microsoft Teams, Jira/Confluence and Monday.com. No doubt you've heard of some of these applications, as many gained in popularity when the world moved to working from home during the COVID-19 pandemic. The goal of these applications is mostly to move individuals away from email-based work, predominantly promoting themselves on the basis that you can be more productive by using their software. Slack describes itself as a place 'where work flows...where the people you need, the information you share, and the tools you use come together to get things done'. It's an odd thought that we might need another piece of software to be the place where we finally get things done. And the use of the phrase 'where work flows' seems contradictory to its very premise. Slack is built on the idea that it allows you easier access to the people you need. But what if those people are busy doing work? What if they are in a state of flow? By using Slack you break that state. You interrupt their work. It makes a fantastic slogan, but Slack is just another messaging application that employees use to communicate with each other.

Zoom is another piece of software that has become widely used as a result of the pandemic. The introduction of the phrase 'I'll Zoom you' into our modern lexicon has been probably the most light-hearted burden placed on society by COVID-19. The popularity of the application exploded during the pandemic. In 2019, the number of meeting participants on Zoom per day was about ten million. By the end of 2020 that number rose to 350 million. The company's US$21 million profit in 2019 was probably the CEO's bonus in 2020, when profits hit US$671 million. Its use has become so common in many companies, that even though people are back in the office, employees will still hold meetings through Zoom rather than bothering to book a conference room. Zoom would no doubt fall into the productivity software category, as it prides itself on its convenience and ability to connect people through video conferencing, therefore allowing work to get done.

However, it is important to look beyond the fancy façade these productivity programs present and examine just how effective they are. To do that we can look at how they are used with data provided by a company called RescueTime. RescueTime monitors what you are doing when you are on your computer and provides you insightful data to help you understand where maybe too much of your time is being spent. It tracks when you are browsing websites that aren't productive and also tracks when you are actually being productive with your device. RescueTime used to analyse the work habits of all their users and produce an annual summary. Back in 2017 they released the results from over 225 million hours of screen usage from their user base. They found that individuals who use productivity tools like Slack spend around ten per cent of their day on these tools alone. So, not only are people spending an hour a day on their phones, they are also spending another fifty minutes of an eight-hour work day sending Slack messages.

The distraction problem plagues workplaces. RescueTime found that on average we use fifty-six different applications and websites per day, jumping between them 300 times every day. In their 2018 annual survey, they noted that users check their emails or Slack on average every six minutes. How can we expect any meaningful work to get done when individuals are not focused on the task at hand for longer than six minutes at a time?

Instead of having multiple different productivity applications to help us complete our jobs and stay constantly connected, perhaps we need software that allows us to disconnect our devices and focus on a task for longer than six minutes. This is where Zenware comes in. Zenware essentially provides disconnection from everything else on your computer except the task at hand, with the goal of helping the user to be more productive. You may be familiar with the focus view on Microsoft Word where the whole computer screen just becomes your Word document and the application creates an environment that allows you to focus on the one document you have opened in front of you. Zenware applications are similar but they tend to go one step further by blocking out certain applications – like web browsers – altogether. With a Zenware application, you can't alt-tab across to a web browser from your focus mode in Word. In some ways they are the antithesis of productivity tools. Whereas productivity tools promote connection, Zenware promotes disconnection. However, both sets of software have the aim of improving how well you work. So, which is better?

The results from users of Zenware tell a very different story to the users of productivity applications who are changing tasks every six minutes. The incredibly successful storyteller George RR Martin has written around 1.7 million words for his hugely popular Game of Thrones series. He has sold more than twenty-five million copies of these books, and Game of Thrones is the most watched show in the history of HBO, the American broadcasting company that turned the books into a television series. So how does George do it? How does he write so many words and tell such brilliant stories? Is he constantly connected to his publisher and editor through a productivity tool, getting instant feedback on his work as he writes? Does he check in periodically on social media and listen to the feedback from fans? Of course he doesn't. Martin uses a program called WordStar 4.0. Being in its fourth version makes it sound fancy, right? Well below is an image that shows just how fancy WordStar 4.0 is:

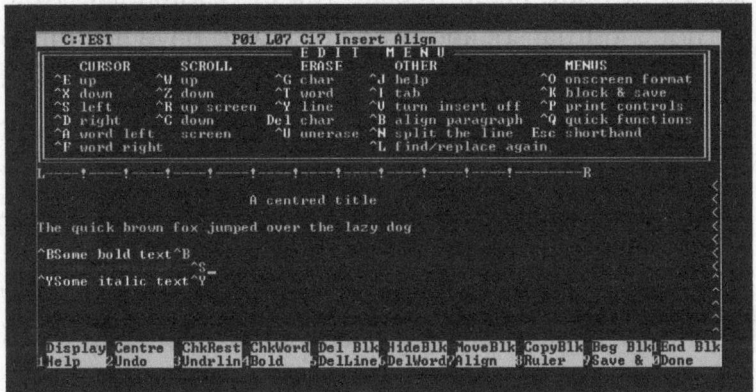

If you can make sense of that image, then bravo. Most of us, if confronted with that on a screen, would give up in less than thirty seconds. Martin says, 'I still do all my writing on an old DOS machine running WordStar 4.0, the Duesenberg of word processing software.' Martin clearly understands that his best work is distraction-free work done in an environment of disconnection rather than connection. His work habits are reflected in the quality of work he produces and the popularity of that work. Without this focused work environment, you could easily argue that Martin would not produce work of such high quality.

But Martin isn't the only individual benefitting from avoiding distraction when they work. Many successful people have incorporated Zenware into their working life so that they, too, can produce their best work. Freedom, which is a Zenware software that allows you to block whichever distracting applications and websites you want, has a website full of testimonials from authors, journalists and academics who have used it to help them achieve more. J.T. Ellison, *New York Times* bestselling author, says that she has written about two million words of fiction and about twice that in nonfiction using Freedom. That amounts to fourteen novels and countless short stories. Paul Guyot, television and film screenwriter of shows and films such as *NCIS: New Orleans*, *The Librarians* and *Judging Amy*, uses Freedom to help stay focused for three- to four-hour blocks of writing. And Scott Cunningham, Associate Professor of Economics at the University of Baylor, has used Freedom to help him overcome distractions and his ADHD and work towards a tenured position.

But it's hard to argue with monetary success. Slack listed on the New York Stock Exchange in June 2019 with the ironic stock ticker WORK. In July 2021 it was purchased by Salesforce for US$27.7 billion, netting its founders unimaginable wealth in a short period of time. Both Zoom and Atlassian, a software company that provides project and content management tools that are typically labelled as productivity tools, have market capitalisations that are well into the tens of billions. These are companies that dwarf the size of some of the oldest businesses in the world, all because they help people work 'better'. Zenware companies like Rescue-Time and Freedom aren't publicly traded companies. They're small companies that exist on the internet and provide their products to a small group of people. If they did trade publicly, they wouldn't achieve close to the value of the more popular productivity applications. But tools that promote deep thinking rather than shallow thinking, that promote disconnection rather than connection, are the tools that are truly able to change how a corporate individual feels about their workday and workplace. They have the power to remove distraction and improve employee happiness by giving them back focus. And by giving them back focus you provide them with the opportunity to have a truly satisfying career.

Hamlet's Blackberry

Zenware is just one of the many ways you can combat distractions in the workplace. It is one of the seven different tools author William Powers promotes in his book *Hamlet's Blackberry* to help readers build a good life in the digital age. Power's goal is to provide the reader with practical tools to help them address the technology overload in their everyday lives. The tools he promotes translate incredibly well into a corporate culture.

Power's first principle is distance. That means finding ways to put space between yourself and the connectedness of technology. For him it's not just about not using technology, it's about creating opportunities to be disconnected from the hyperconnected world we live in. He recounts a story from Plato that details a dialogue between Socrates and his friend Phaedrus. Amongst other matters, Socrates takes issue with Phaedrus and his use of the latest form of communication technology in those times, written word. Socrates is adamant that the written word will inhibit people's use of their memory, making them forgetful and lazy. To deal with this, the two take a walk outside the walls of Athens into the countryside to discuss the matter. They create physical distance between themselves and the crowd and find time to be away from it all, to contemplate the matter, to truly think about the issue at hand. Powers gives us the modern equivalent:

Physical distance is the oldest method of crowd control. In one obvious sense, today it's much harder to go outside the "walls" of the connected life. Truly disconnected places are increasingly rare. But in another way, it's easier. Take a walk without a digital gadget, and distance is yours. The moment you leave all screens behind, you're outside the walls.

The rarity of this event, though, speaks to the problem at hand. We almost never go out and deliberately leave our mobile phone at home. It would be irresponsible to be unreachable in case of an emergency. But let's really consider how often we have taken our phone with us on an outing and it has saved our life or a disaster would have occurred if we had been unreachable for fifteen minutes. I cannot personally think of any circumstance in my life where it's been vital that I've had my phone on me. It's been convenient, but not lifesaving. But, as Powers points out, it is hard to give up:

Though a smart phone brings convenience and a sense of security, it takes away the possibility of true separateness. It's a psychic leash, and the mind can feel it tugging. That's the problem: we've gotten so used to the tug, it's hard to imagine life without it.

Putting physical distance between yourself and your digital devices is an opportunity to fight against the mental tug. It doesn't have to be all day; it can just be small portions of your day where you refuse to look at screens. Of course, if you are going on a solo bush-hike for the first time by yourself, take your phone. But if you are taking your dog for a walk in the park, consider leaving it behind.

This is a tactic I employ personally. I make sure I avoid screens for as long as I can in the morning. This doesn't mean I might not do something productive with a screen, but my phone stays in another room and I don't open any email applications or web browsers. I actually hate the idea of being on the train and looking at my emails in the morning. For me, the morning is the most productive part of the day. It's important to understand who you give access to the most productive part of your day and who's in control of that time. It should be you; it shouldn't be distraction. Putting physical distance between yourself and your distractions is one of the easiest ways for you to overcome them, and by doing so you give yourself opportunities to take control over the most productive part of your day, whenever that may be.

Escaping the digital world can be difficult. And sometimes even a ten minute phone-free stroll isn't enough to quiet the noise of the distraction in our lives. But finding ways to deal with distraction when we are sitting inches from a computer screen is necessary for coping with the digital world we inhabit. Powers sees the need to find 'inner space' as an issue that's plagued society since ancient times, and he draws on the writings of the ancient Roman philosopher Seneca to demonstrate this. Lucius Annaeus Seneca was born sometime around 4 BC. He was heavily involved in politics as a Roman senator but is best known for his writings as a philosopher. Seneca lived in a crowded world, and his writings speak of the difficulties he faced navigating this constant noise. It's important for us not to fall into the trap of comparing our era as significantly busier than any other previously. It's easy for us to think how focused and undistracted we might have been had we lived and worked 50, 100 or even 2000 years ago. But it's about the relative experience of that individual at that point in history. For Seneca, the world was busier than any other time period that had ever been before, and dealing with this was a real issue he felt he needed to work through. Seneca directly addresses the need to take time away from the crowd in his writings to an old friend named Lucilius:

You ask me to say what you should consider it particularly important to avoid. My answer is this: a mass crowd. It is something to which you cannot entrust yourself yet without risk ... I never come back home with quite the same moral character I went out with; something or other becomes unsettled where I had achieved internal peace.

All the themes Seneca addresses here are things we can relate to. Our experience with technology is continually about connecting with the crowd, and when we connect with the crowd constantly, something starts to feel off. We come back from our experiences of connection a bit more unsettled than we were when we started. We feel like something has changed. The crowd takes away opportunities to achieve the 'internal peace' Seneca talks of by introducing a mass of noise into our everyday life. Without time away from this noise, finding a sense of being settled in our environment is extremely difficult.

So how does Seneca avoid the crowd? Seneca advises his friend Lucilius to 'measure your life: it just does not have room for so much.' For Seneca finding that internal peace again was a matter of simplifying his life and his thoughts. 'After running over a lot of different thoughts, pick out one to be digested thoroughly that day', advises Seneca.

A lot of the distractions in our lives are often self-imposed. Do we really need to check our inbox every six minutes? Do we really need to have twenty different tabs open on our web browser? Must we be constantly available for our colleagues or clients? It's a matter of approaching technology in a meaningful way that drives thoughtful interaction. Understanding where to draw the line with your device is the most important part of how you use it. We often feel overwhelmed because we have tried to do too much and being constantly accessible does not help the cause. Consider how you might simplify your interactions with your device and how this might help you achieve the 'internal peace' that Seneca so highly valued.

Away and Disconnected

At the company run by the 'onwards and upwards' manager, we all had the messaging application Skype on our laptops. Skype tries to show your colleagues what you are doing at work. If you are at your computer sending emails or writing in a document, your status is shown as 'Active' and a pleasant green icon appears next to your name in Skype. Being online implied you were doing work and were available to respond in an instant to other people if they were to message you. If you were in a meeting, a serious red icon was shown to indicate to your colleagues that you have important matters to attend to and your focus is elsewhere. But if you were to lock your computer or walk away for long enough, the computer would go to sleep and the status would change to 'Away' which appeared as a passive aggressive yellow icon. The colours of the symbols were enough to indicate that if we were online and connected, we were doing work and being good, and if we were away and disconnected, we weren't. Away meant we weren't doing work.

Whilst at this company, I obtained the privilege of working from home. But I often had only a couple of hours of work to do, so I had to get creative in keeping the green icon next to my name. In order to stay online and connected, I would place a wireless mouse in my pocket. As I moved around the house or bounced my leg when I sat down, the mouse would move on the screen and the computer would think I was still sitting there doing work. The Skype application forced us to be connected, and I satisfied that requirement.

This is the reality of the corporate environment many of us inhabit. If you are connected, you are doing work.

If you are disconnected, you are being lazy. But the reality is, if you are connected you are often distracted, and when you are distracted, you aren't productive. Also, when you are distracted, you are often dissatisfied with your career and your job. You go home and wonder what you spent your day doing. Disconnecting is a viable and, in fact, better alternative to being connected when it comes to being productive.

To improve the quality of our careers, we don't need more so-called productivity applications to fix the problems already caused by other pieces of software. It's like trying to fix the drinking problems of an alcoholic with a different kind of alcoholic beverage – 'Have you tried gin?' Instead, we need to adopt habits such as putting physical distance between ourselves and our distractions. We need to simplify our interactions with technology and change how we work and engage with screens by taking a more deliberate approach. By doing this we will be able to disconnect ourselves from our environment of distraction and give ourselves the space we need to take back our careers.

Chapter 4

The Loss of Leadership

*All of humanity's problems stem from man's
inability to sit quietly in a room alone.*
Blaise Pascal

Leadership is one of the most overused and abused words in the English language today. It's a topic that never fails to see airtime on a news channel. If you scroll through a social media feed, you are almost guaranteed to find a post about leadership with the #leadership conveniently attached underneath. Something along the lines of 'Top ten things successful leaders do in the morning' is a mainstay article published on LinkedIn. Questions of leadership constantly permeate our society: What makes a great leader? Why would people follow you? How do you become a better leader? The reality of these kinds of questions is that you could search the entire internet and never find a concrete answer. Leadership is one of the most undefinable concepts that exists in our society today. Saying a person is a good leader has the same weight as saying that person is cool. The reality is all sorts of people are labelled as cool, but no one can ever tell you what it actually means.

Although great leadership is hard to define, like 'cool' we know it when we see it. One approach to better understand great leadership is to study what great leaders do and try to understand what makes them different to us and other leaders. By doing this we can uncover one of their most important and underappreciated attributes - their ability to sit quietly in a room, alone.

'You're Fired'

It's almost impossible to talk about leadership without talking about politicians and businesspeople. When we place in our mind the image of a leader, invariably we tend to think of the leader of a country or the CEO of a large enterprise – someone who has risen to the top. One of the more surprising individuals to rise to the top of both a powerful country and a large corporate structure is Donald Trump. From 2015 to the start of 2021, he was arguably the world's most talked about political figure. He has been involved in more controversies than any previous US president and has been under the microscope since the start of his controversial election campaign. Whilst being constantly examined, he has managed to escape some of the most damning reports imaginable: from paying a porn star to not leak details of his extramarital affair, to having articles of impeachment brought against him over his dealings with the Ukrainian government and his influence on the rioters who attacked the United States Capitol building in Washington DC.

Prior to COVID-19, it looked almost certain he would be re-elected, in spite of all these shortcomings. But failing to address the needs of his voters during the greatest pandemic of our lives was finally his undoing. Only just, though, as he still amassed over seventy-four million votes in the popular vote. Seventy-four million people either voted against Joe Biden or for Donald Trump. It's an impressive feat when you consider how closely scrutinised he was during the four years he was president. To survive his many controversies and scandals, Trump relied on the age-old art of lying, or at least avoiding the truth. The *Washington Post* fact checker recorded Donald Trump making 30,573 false or misleading claims over four years (1461 days). That's about twenty-one lies every day, which works out to be one lie every sixty-eight minutes. He has constantly denied any wrongdoing and has escaped relatively unharmed. How is this man, someone who has unquestionably done the wrong thing multiple times and lied openly about it, still a leader?

There is, in my mind, a rational explanation to this. Trump is excellent at appealing to a distracted population. He is the first president to be in office and have a regular, active, personal presence on Twitter. But really the phrase 'regular, active, personal presence on Twitter' doesn't come close to describing his usage levels. His Twitter statistics are extraordinary. From the time he announced his presidential campaign in June 2015 until March 2020, Trump tweeted 17,319 times. That's ten tweets a day. Over eighteen waking hours, that's one tweet every 107 minutes – a piece of information sent out to people every two hours for five years straight, non-stop. If you did something every two hours for five years without a break, you'd be amazing at it too.

The way politicians connect with the voting population has changed just as dramatically as the way the voting population connects with each other. Fifteen years ago, an internet presence for a politician was probably a basic website that outlined some of the key points of their policies. They maintained the typical methods of reaching the voting population by holding rallies and ousting their opponents during presidential debates. They'd use television, radio and newspapers to advertise. But none of those media have the reach that social media does. Now politicians can reach millions of people every two hours for five years straight. Trump could reach seventy-three million people, all of his Twitter followers, multiple times every day. That power dwarfs the capabilities of all the alternative media put together. It's this reach, this constant stream of information he poured into his supporters, that enabled him to maintain his status as leader. And it's the frequency of his online activity that provides insight into his reality. This man arguably defines the term distracted. I can't imagine how much would have been running through his mind as he juggled constant controversy, global politics, a pandemic and a Twitter addiction. This man's mind must have barely had time to stop at all during his presidency. He was, by all definitions, a distracted president. A distracted president who was victim to all the same pitfalls of the screen as his distracted population. Author and speaker Simon Sinek said of Trump that politicians are a reflection of the voting population – 'a narcissistic population gets narcissistic politicians'. In this case, a distracted population got a distracted president.

One of the best descriptions I can attach to Trump is found in the novel *Heart of Darkness*. If you aren't familiar with the novel, it was written in 1899 and follows the character Marlow as he journeys up the Congo River, taking a job at an ivory trading company as a riverboat captain. He journeys upriver visiting three stations, with the second station, the Central Station, being his longest stop. When he arrives at the Central Station, he meets the manager, who is quite a distinctive character. Below is the excerpt from the novel of Marlow describing the manager of the Central Station. Pay careful attention to the words used to describe him:

> *He was commonplace in complexion, in features, in manners, and in voice. He was of middle size and of ordinary build. His eyes, of the usual blue, were perhaps remarkably cold ... Otherwise there was only an indefinable, faint expression of his lips, something stealthy—a smile—not a smile—I remember it, but I can't explain ... He was a common trader, from his youth up employed in these parts—nothing more. He was obeyed, yet he inspired neither love nor fear, nor even respect. He inspired uneasiness. That was it! Uneasiness. Not a definite mistrust —just uneasiness—nothing more. You have no idea how effective such a ... a ... faculty can be. He had no genius for organizing, for initiative, or for order even ... He had no learning, and no intelligence. His position had come to him—why? ... He originated nothing; he could keep the routine going—that's all. But he was great. He was great by this little thing that it was impossible to tell what could control such a man. He never gave that secret away. Perhaps there was nothing within him. Such a suspicion made one pause.*

William Deresiewicz, American author, essayist and literary critic, used this extract in a lecture delivered at the United States Military Academy at West Point and later published as an essay titled *Solitude and Leadership* to describe some of the leaders he was working with during his university career. He calls attention to the language, noting that Marlow is using terms such as 'commonplace', 'ordinary', 'usual' and 'common' to describe the manager of the Central Station. There was nothing special about that man. Nothing that set him apart as a leader. He had come to his position not because he had excelled at some quality that distinguished him as a leader, but because he could 'keep the routine going'.

In business, this is often the main factor that gets you into a leadership role; if you've served long enough in your junior role you automatically qualify for a position that involves leading others. Intuitively, though, that doesn't really make sense. Length of service doesn't qualify you to lead someone. Leadership is not defined by how experienced you are. In fact, leadership should not be defined by any quality other than itself. Leadership isn't some measure of intelligence, achievement or length of service. It has to be a separate characteristic that is definable in and of itself. But it often isn't. When this happens, it gives rise to leaders such as the commander of the Central Station and Donald Trump – leaders who are obeyed yet inspire 'neither love nor fear, nor even respect,' leaders who inspire 'uneasiness'.

'Uneasiness' is the perfect word to sum up the leadership of Donald Trump – uneasiness inspired by what he says, how he reacts to crises and his overall unpredictability as a leader. You couldn't help but feel uneasy when he addressed the nation about the options to prevent COVID-19 and suggested sunlight and ingesting bleach, later excusing himself from the comments because he was being sarcastic. And although Donald Trump is an extreme example, he represents the uneasiness of distraction in our lives – the uneasiness of always being drawn to a screen, the uneasiness of missing what we truly value and the uneasiness of knowing we are being led by individuals who are just as distracted as we are.

The Soldiers' Home

We turn from one president, Donald Trump, to another, Abraham Lincoln. Lincoln is arguably the most famous American president to have ever held office. He holds this position because he led America during its civil war, one of the toughest periods in the history of the country. The American Civil War started just one month after he took office, and he was thrown right into the deep end. In a letter to his friend Senator Orville Browning, Lincoln wrote that 'the first thing that was handed to me after I entered this room, when I came from the Inauguration, was the letter from Major Anderson saying that their provisions would be exhausted'. Major Anderson was the commanding officer at Fort Sumter, which was, at the time, a key point of conflict that would shape the course of the American Civil War. Right away, Lincoln had to make important decisions about resource allocation and war strategy. His leadership was tested from the very moment he took office.

During this time, not only did he successfully lead the country through civil war, but, in 1863, he also introduced the Emancipation Proclamation. This proclamation, which ultimately became the thirteenth amendment to the US constitution, changed how federal law treated the millions of slaves in the confederate states. This proclamation declared that they, the slaves, 'shall be then, thenceforth and forever free.' The Emancipation Proclamation changed the course of the American Civil War. It shifted from being a fight to contain the rebellious confederate states, to a fight dedicated to the freeing of slaves. What's interesting here is the change of mindset that this introduced: rallying the soldiers behind a cause they were fighting for, freeing slaves. They went from being against something, to being for something. This is a subtle trick of great leadership – the ability to get people to rally behind a cause.

Simon Sinek, in his book *The Infinite Game*, speaks about leaders who lead as if they were playing an infinite game: a game where the rules aren't set, you don't know who else is playing and you can't win. The object of the game is to keep the game going. A great example of an infinite game is business. No one ever actually wins at business. You always wake up the next day to some other problem to overcome or some other area of your business to grow. Business is an infinite game. A finite game is the opposite to an infinite game: the rules are set, you know who's playing and you play to win. Sinek argues that leaders need to know what game they are playing in order to be effective leaders. Leaders of countries play an infinite game. There is no winning politics in the ultimate sense that all the problems end. Every day a leader of a country wakes up to another problem that must be solved. There is no end to the game.

Lincoln was playing an infinite game and he knew how to lead in it. One key element of what Sinek calls an infinite leader is their ability to rally people behind a just cause. For Sinek, a just cause must meet five criteria. The most crucial criteria is that the cause must be for something, rather than against something, which is exactly what the shift in focus for civil war soldiers achieved. It took the fight from being against the confederacy to being for the freeing of slaves. According to Sinek, a just cause must also be:

- inclusive – open to anyone who wants to join
- service orientated – for other people
- resilient – able to endure political, technological and cultural change
- idealistic – big, bold and ultimately unachievable.

Freeing slaves aligns closely with Sinek's definition of a just cause. It was a cause that was open to anyone who believed that slaves should be free. It was something entirely orientated around others. It was an outcome that felt unattainable, but the fight had existed, in one form or another, from the moment slaves were first taken to America. And although we've come a long way in the fight against slavery, it still persists in small pockets of the world. Lincoln was a leader playing the infinite game. He rallied his troops behind a just cause, a cause that meant more to them than just winning a war.

Lincoln is also known for the Gettysburg Address, a speech he made at the dedication of the Soldiers' National Cemetery in Gettysburg, Pennsylvania. Gettysburg had been the site of a horrific battle only four and a half months earlier. Over 50,000 lives were lost and many thousands more were injured. Lincoln spoke for only two minutes, during which he said some of the most famous words ever delivered by an American president.

Four score and seven years ago our fathers brought forth on this continent, a new nation, conceived in Liberty, and dedicated to the proposition that all men are created equal.

Now we are engaged in a great civil war, testing whether that nation, or any nation so conceived and so dedicated, can long endure. We are met on a great battle-field of that war. We have come to dedicate a portion of that field, as a final resting place for those who here gave their lives that that nation might live. It is altogether fitting and proper that we should do this.

But, in a larger sense, we can not dedicate -- we can not consecrate -- we can not hallow -- this ground. The brave men, living and dead, who struggled here, have consecrated it, far above our poor power to add or detract. The world will little note, nor long remember what we say here, but it can never forget what they did here. It is for us the living, rather, to be dedicated here to the unfinished work which they who fought here have thus far so nobly advanced. It is rather for us to be here dedicated to the great task remaining before us -- that from these honored dead we take increased devotion to that cause for which they gave the last full measure of devotion -- that we here highly resolve that these dead shall not have died in vain -- that this nation, under God, shall have a new birth of freedom -- and that government of the people, by the people, for the people, shall not perish from the earth.

You couldn't have picked a tougher location or harder moment in which to unite people around a just cause. Lincoln addressed 15,000 people that day, and he needed to have all those present look beyond the reality that death was a real consequence of pursuing the freedom of slaves. He did this by turning the attention of the audience not on himself but on the people who gave their lives fighting for the just cause that everyone there believed in. He knew this wasn't some political fight to win, or a chance to get more votes in the next election. This wasn't an opportunity to demonstrate how much better the Union was than the Confederacy. This wasn't an opportunity to let his voting public know that 'our thoughts and prayers are with those who have died and their families'. This was an opportunity for a great leader to pay tribute to those who died by advancing the very cause they sacrificed themselves for. The eloquence and grace of this speech is clearly from the mind of a leader we haven't seen for some time.

Lincoln was only in office for five years before he was assassinated, in 1865. In that time, however, he navigated some of the most complex situations in the history of America and gave one of the greatest speeches ever. We would all be hard pressed to think of a leader or politician in our lives who has made such an impact in such a short period of time.

It's important to realise that Lincoln was a human being, just like us. It is easy to reflect on these historical figures and see them as different to us. It is a natural assumption to make, given all Lincoln did in his life. We make out that it is impossible to have these kinds of people exist today. But the reality is, if Lincoln had lived in today's world, there is a good chance he would have achieved great things in this period of history as well.

There is an old saying I often repeat to myself that goes: 'Thoughts produce actions, actions produce habits, habits produce lifestyles and lifestyles produce destinies.' Lincoln wasn't a man destined to achieve greatness from birth. His life was built, first and foremost, on a foundation of being a thinker which is what ultimately led him to achieve greatness. If we can understand what Lincoln did to shape his thoughts, we can uncover what he did to shape his destiny.

Lincoln spent a lot of time at the White House, as you would expect. Whilst there, he was often surrounded by individuals trying to get his attention and time. Renowned Lincoln scholar Harold Holzer says that 'virtually from Lincoln's first day in office, a crush of visitors besieged the White House stairways and corridors, climbed through windows at levees, and camped outside Lincoln's office door'. It is in this environment that we see a key habit of Lincoln's that indicates how he was able to shape his destiny and become a true leader. From 1862 to 1864, during each summer and early autumn, Lincoln would commute, on horseback, between the White House and the Soldiers' Home. It wasn't the safest of journeys, with Lincoln requiring accompaniment by soldiers and, on more than one occasion, being attacked. But the Soldiers' Home was an important place for Lincoln, so he saw great value in travelling there, despite the risk to his life. Originally built in 1851, the Soldiers'

Home was a place for disabled veterans of war. There, Lincoln found an opportunity to experience that which he simply could not experience at the White House. Lincoln found a chance to be alone with his thoughts. Although he was attended to by his household staff at the Soldiers' Home and there were two companies of volunteer soldiers camped out on the lawn to help keep him safe, Lincoln used his time there to think. We know this from the many accounts of people coming to visit Lincoln at the Soldiers Home. Visitors would often find him entirely alone in a room, deep in thought. In the weeks leading up to the Gettysburg address, he was often found walking alone at night in the military cemetery on the premises. Not only was this a fitting location to reflect on the sacrifice made by soldiers, it was also the perfect spot for Lincoln to be undisturbed while thinking.

Spending time alone with his thoughts was one of the habits that formed Lincoln into the man he was. It was a habit that shaped his destiny. Lincoln had what Trump, or any distracted politician, never has. He had time alone with his thoughts. He had a chance to really think about the issues facing the nation he was leading and how his decisions would affect the people who were relying on him – without disruption from others, without noise from the outside world. It is this habit that I believe set Lincoln apart as a leader, and it's the absence of this habit that is the reason why so many of our leaders today fail at what they do. They, just like most of the voting population, rarely make time to think in isolation.

Isolation Is a Gift

The term isolation was thrown into the forefront of our collective consciousness during the COVID-19 pandemic. Never before had humanity had a term so quickly thrust upon them, so it would be difficult to consider it without first looking at how it was thought about and dealt with during the pandemic. Isolation, or iso as it is often referred to, was experienced by most people throughout the COVID-19 pandemic, either by government order or personal choice. To reduce the risk of infection, people were forced to be separated from others. However, during this time we certainly weren't forced to be disconnected from people. When you look at advertisements and messaging from public figures during the pandemic, there were constant encouragements to stay connected to others through digital platforms. We were again pushed the idea that disconnection is bad and connection is good. COVID-19 is certainly an extreme scenario, and it would be impossible to defend the position that the connection by digital means during the pandemic wasn't beneficial. I'm sure many of you were very thankful for being able to stay connected during this time through various digital tools. Personally, I found that my best connections with people were through good old-fashioned phone calls. But because of this constant ability to be connected we didn't really experience isolation in its truest form.

The companies that facilitated this connection experienced massive growth during the pandemic. In the space of six months during 2020, Apple went from being the first US$1 trillion company to the first US$2 trillion company, as the demand for its products, products that offer constant connection, soared. Facebook's share price reached all-time highs during the pandemic, as more people than ever before logged in to the platform. 'People want to stay connected while being asked to maintain social distancing and eliminate loneliness,' said Facebook CEO Mark Zuckerberg during a conference call. Another way people chose to navigate isolation during the pandemic was through video streaming services. Netflix added sixteen million new subscribers in one quarter, which was about a ten per cent growth in a company that usually only adds a few million new subscribers every few months. When people were forced into isolation by the pandemic, they made more calls on video chat, spent more time on Facebook, watched more Netflix and walked their dogs more frequently than ever before. Clearly our experience of isolation and Abraham Lincoln's were very different. We couldn't have tried harder to find ways around it, but Lincoln risked his life to have time alone.

How do we make sense of these two very different reactions? Let us start by understanding what isolation allows us to experience – solitude. German American philosopher Paul Tillich summarised solitude by saying it 'expresses the glory of being alone'. Solitude is what we experience when we are given an opportunity to be alone with our thoughts. Solitude provides us with a chance to, without any noise or input, understand what we truly think or how we truly feel about an issue. In other words, solitude is the antithesis of the internet. At the Soldiers' Home, Lincoln experienced solitude in its finest form. He was able to think uninterrupted for long periods, not just in short intervals between an urgent request or a meeting or a Tweet.

The screens and distractions in our lives prevent us from experiencing solitude, from being truly alone with our thoughts. Although this chapter is focused on solitude as a requirement for leadership, I would go so far as to say that true solitude can enhance anyone's experience of life. It lets us stop and ask scary questions that we often avoid asking ourselves. Am I happy? Do I believe in what I do every day? Do I feel as if I am making a difference? Often, we don't want to hear the answers to those questions, so we block out our own thoughts with the noise of our Instagram feeds and run away from them as we plunge down another YouTube rabbit hole. But solitude allows us to take a moment to ask those questions. Chances are we won't find the real answer the first time we ask these questions. But solitude allows mulling over to occur, the refining of ideas, that ultimately progresses us towards an answer.

Answers to life's big questions are not found in the noise or in the distractions in our lives. They are found in solitude. Solitude allows you to really get to know yourself, to find out what you want, what you believe in, what you value and to ask yourself if what you are doing aligns with those things. Solitude even allows you the space to tell yourself that you were wrong. Imagine a world where the reaction of the people around us (including our leaders) wasn't defensiveness when confronted over a wrong decision, it was honesty. Imagine a world where the first reaction of people was to take responsibility and promise to try to do better next time, not to make an excuse as to why they might still be right. We will never get this world from each other until solitude is a part of our lives, and we will never elect leaders that value solitude until we value it ourselves.

You can find solitude by isolating yourself, by simply sitting alone in a room with no distractions, just as Lincoln did. You can also achieve solitude in other ways. William Deresiewicz, in *Solitude and Leadership*, talks about three other strategies for finding solitude. The first one is exactly what you are doing now – reading books. Why books and not blog posts, wall posts or Tweets? What's written online is often produced in a distracted environment, lacks depth and is typically written in small, bite-sized pieces for a distracted audience. Reading a book is very different. Deresiewicz argues that there are two advantages to books. One is that they are often the result of someone else's solitude, someone else's time alone with their thoughts, their attempt to think for themselves. The other advantage is that a lot of books are old. That means their ideas have stood the test of time, that their ideas are valuable and transcend the culture they were originally intended for.

Reading books also encourages us to pause and reflect on what's being written. When online, we jump from one piece of information to the next. Often we never actually read every word written on the page. We take snappy headlines and proclaim them as fact. We scan articles looking for quotes that are easy to digest or are given by people we believe are intelligent, in the hope that if we can requote them to others, we too will obtain a similar level of respect. But when we take in information like this, we retain little of it, and what we retain is not our own. Reading books forces us into understanding the intricacies of someone's arguments, the intricacies of thoughts produced in solitude. This in turn sharpens our thoughts, and we emerge from our time reading not merely with more information, but with more knowledge – knowledge not only about how the world itself works, but how our personal world works too.

Deresiewicz also argues that physical labour is a form of solitude. He quotes another excerpt from Heart of Darkness. Marlow has just finished speaking to the second-in-command at the Central Station, an individual whose description we will touch on later. He describes what happens as he walks away from that conversation:

It was a great comfort to turn from that chap to ... the battered, twisted, ruined, tin-pot steamboat ... I had expended enough hard work on her to make me love her. No influential friend would have served me better. She had given me a chance to come out a bit—to find out what I could do. No, I don't like work. I had rather laze about and think of all the fine things that can be done. I don't like work—no man does—but I like what is in the work—the chance to find yourself. Your own reality—for yourself, not for others—what no other man can ever know.

Marlow makes a fine point here, one that we know all too well and experience every day. Work is hard. It's evidenced in our earliest stories and has formed a crucial part of our narrative. In the Genesis account of the fall of man, God places the burden of work onto mankind as a consequence for their actions: 'By the sweat of your face you shall eat bread.' The reality that physical work is difficult seems intuitive, but Marlow points this out to call attention to what comes of one's attempts to pursue difficult work, particularly physical work – there is something beautiful to be found in it. There is something therapeutic about it. Being someone who works a desk job, I often think about jobs that are predominantly physical, jobs where you get out every day and build or create something with your hands. There must be something more satisfying about building a table than there is about writing an email.

When we purchased our house it took me a while to appreciate how much work goes into maintaining a home. One of the main jobs I hadn't considered was mowing the lawns. It's a job that needs to be done every couple of weeks when you are a homeowner. But after moving in I was determined to join the collective of Australian men who proudly do their lawns. To do this I needed equipment. So I ventured into the local Bunnings hardware warehouse – a store familiar to many Australians. Bunnings is a massive warehouse of everything you would need to take care of your home or garden. A maze of goods that seems to never end. So every time I walk into a Bunnings store, I feel as if I need to get a special sign attached to my t-shirt that says 'I work in finance, please help me'. Nevertheless, I went to the lawn mowing aisle with the determination befitting a man who is ready to do his house and street proud. I quickly picked out my blue lawnmower and matching blue whipper snipper, and I even remembered to pick up motor oil. But my determination faded when I had to find a whipper snipper line. I must have stared at the wall of options for about ten minutes. I simply couldn't figure out which line was best for my new blue whipper snipper. It felt like I was doing a test, and it was the simple question that had stopped me in my tracks.

Thankfully, a man saw the difficulty I was having and offered to give me a hand. He was a lawnmower savant and quickly corrected all my mistakes, changing my whipper snipper over for one that actually had line I could buy and telling me all the things to do to avoid disaster. Several hundred dollars later my car was full of mechanical wonders ready to assist me in my manly duty of mowing the lawn.

Being on a hilly block of land means that it is a fair bit of work to mow our lawns. It takes me about three hours to get the job done. My wife often insists we should pay someone to do it for us and give me back a few hours on my weekend. But I refuse. After a week of emails and Zoom calls, I love the physical work of mowing the lawns.

In his classic book on solitude, *Walden*, Henry David Thoreau spends countless pages recounting in great detail his experience living alone in the woods in what he describes as his attempt to learn to 'live deliberately'. An entire chapter of *Walden* is devoted to his experience tending to his beanfield. In all, Thoreau planted over seven miles of beans. It was an entirely impractical amount for one man to manage. He recalls that the first beans were ready to be harvested before the last were even planted:

What was the meaning of this small Herculean labour, I knew not. I came to love my rows, my beans, though so many more than I wanted ... What shall I learn of beans or beans of me?

Thoreau's statement about his field of beans is a reflection of the truth of manual labour: in it there is something to be learnt about oneself. French philosopher Alexandre Kojève puts it like this:

The man who works recognises his own product in the World that has actually been transformed by his work; he recognises himself in it, he sees in it his own human reality, in it he discovers and reveals to others the objective reality of his humanity, of the originally abstract and purely subjective idea that he has of himself.

Manual work gives us a renewed sense of what we are capable of because we are able to manifest our efforts in something external to ourselves. Our blood, sweat and tears are poured into a project, however great or small, and we see the best and worst of ourselves in it. And the solitude of manual labour enables a deeper knowledge of who we are.

Each time I mow the lawns I get to see what I'm capable of. And each time I see a chance for improvement, and so the next time they are a little better. I realised this is a reflection of the rest of my life: I try and get a little better at most things, but often see room for improvement. Just as Marlow found himself in fixing a ship and Thoreau learnt from his field of beans, I too emerge each time from mowing the lawns understanding a little more about myself.

The last form of solitude that Deresiewicz explores is perhaps the least intuitive because it doesn't involve being alone. It is the solitude of deep and meaningful conversation with those closest to us in our lives. Ralph Waldo Emerson described it best when he said that 'the soul environs itself with friends, that it may enter into a grander self-acquaintance or solitude'. For Emerson, 'friends' carries a different meaning than many people currently experience. The friends you can experience solitude with are those you can trust with anything. These are the people who will hear your deepest secrets and help you guard them. They will call you out when you are lying to yourself. And they will actually be there when you need them. These are the people who will seem more interested in the conversation you are having with them than anything else that is happening around them. They will look at you and not their phone when you talk to them. Friends for Emerson are not all the people who follow you on Instagram, Facebook or Twitter, or who are a part of your Saturday social WhatsApp group. These are not the people who religiously like your posts on social media. Interacting with friends through screens isn't a form of solitude, it's simply, as William Deresiewicz points out, another form of distraction:

> *Instead of having one or two true friends that we can sit and talk to for three hours at a time, we have 968 'friends' that we never actually talk to; instead we just bounce one-line messages off them a hundred times a day. This is not friendship, this is distraction.*

There might be only a few people in your life who really fit the description of someone with whom you can experience solitude. Perhaps none of your friends fit that description. If that is the case, I would encourage you to rethink your definition of friendship, rethink your interactions with people and what you are doing to maintain relationships with friends and family. We would all most likely say that we value our friends and family, but what do we do in our lives to demonstrate that value? We stay 'connected' through a family messaging group or a Facebook group, and we ensure that any life event is well received through social media: that birthday posts are made on people's walls and that baby pictures are given an adoring heart emoji rather than a like button. For many, if not most, this is the extent of our friendship.

If you have ever been so deep in discussion with one or two friends that the rest of the world ceased to exist, you will understand the solitude of friendship. So, instead of being distracted by 968 'friends', focus on those one or two who will help you experience solitude. Focus on the friends who will help you to get to know yourself in a way that would otherwise be impossible.

We need to make solitude, whether it be a time of personal introspection, reading, physical labour or deep, intimate conversation, part of our life. And this is particularly important for our leaders. Imagine a leader who took time to be alone and think without distractions, a leader who spent time immersed in books, a leader who created things with their hands and who had deep, intimate conversations with others. Imagine the kind of thinker they would be. They would be able to beautifully articulate positions on issues. They would be open and receptive to others. And they would understand that there is no such thing as winning in leadership. This kind of leader wouldn't be on LinkedIn posting about the top ten steps to becoming a great leader. They wouldn't be scrolling through their social media feed during a team meeting. And they wouldn't be sitting in the Oval Office tweeting.

A Brief History of Walking

During the pandemic, going for walks was one of the main ways that allowed an escape from lockdown. It was also the only real way you were allowed to connect with friends or family in person. Often you would see people walking and talking together, enjoying the solitude that comes from deep conversation. You must wonder how many of the world's biggest problems were solved during those walks. Walking is the final form of solitude we will consider as it provides opportunities for both conversation like that which Emerson talked about, but also personal introspection when walking alone. Unfortunately, technology often hijacks this form of solitude. People dare not leave their phone at home and often when walking people are either looking at their phones, listening to their headphones, or being constantly interrupted on their journey by a notification from their device.

The act of walking has long been used by many individuals as a way to help process their thoughts and get away from the noise. In his book, *A Philosophy of Walking*, Frédéric Gros provides an analysis of the finer details of what it means to be a walker. In the chapter titled *Solitude*, Gros explains how even though you are alone physically when you are walking, you are not truly alone:

Lastly, you are not alone because when you walk, you soon become two. Especially after walking for a long time.

What I mean is that even when I am alone, there is always this dialogue between the body and the soul. When the walking is steady and continuous, I encourage, praise, congratulate: good legs, carrying me along ... almost patting my thigh, as one pats the withers of a horse ... When I walk, I soon become two. My body and me: a couple, an old story.

This notion of walking as a form of solitude echoes the concept of solitude being experienced with two or more people. But here, whilst walking, it is you and your soul. You and the dialogue in your mind. You process ideas, solve problems and justify ideas in order that when you return from your walk you might better know your way around your own life. These benefits from the solitude experienced whilst walking are evident in the lives of many prominent individuals throughout history. In his book Gros draws on examples of great minds who incorporated walking as part of their lifestyles and shows how walking helped to shape them into the people they became.

Charles Darwin was an avid walker. There was a gravel track not far from his home known as the Sandwalk. Darwin would walk this track twice a day, every day, rain or shine, in sickness or health. During his walks he would think through problems and uncover solutions that he could not think of whilst staying seated. Walking shaped Darwin into the person, and thinker, that he became.

Jesus of Nazareth spent the majority of his three years of ministry walking from city to city. He made a name for himself by telling parables – stories with a message. These were stories inspired by years of study of Jewish literature. This time spent walking, either alone or with his disciples, would have allowed him to shape and refine these stories that were fundamental to his ministry.

Walking was an integral part of Gandhi's success. His daily walks gave him time to think through his position and refine his approach time and time again. Gros argues that walking embodied many parts of Gandhi's approach to his goal of liberation from the British empire. It aligned with his themes of simplification, you only walk with what is necessary, autonomy, independence from the British, and firmness and endurance, knowing that the road to independence was long and hard. He inspired many to join his cause through his walks, and there is no doubt he won many over to his side through deep conversation during this time.

William Wordsworth, the great English poet who laid the foundation for the Romantic Age in English literature, once estimated that he had covered between 175 and 180 thousand miles during a lifetime of walking. Gros argues that Wordsworth was the first 'to conceive of the walk as a poetic act, a communion with Nature, fulfilment of the body, contemplation of the landscape.'

Gros notes that when Wordworth's sister was asked where her brother worked, 'she waved vaguely at the garden and said 'that's his office." There is no doubt that Wordsworth produced his influential writing only as a result of the internal dialogue that transpired throughout his walks.

Jean Jacques Rousseau, French philosopher and writer, also saw walking as key to his success as a thinker. Gros says that for Rousseau, 'it was paths that stimulated his imagination.' Rousseau almost couldn't think without walking, saying:

There is something about walking which stimulates and enlivens my thoughts. When I stay in one place I can hardly think at all; my body has to be on the move to set my mind going.

Thoreau covers walking as a place and time for solitude in both *Walden* and the aptly titled *Walking*. Thoreau used walking as a way to protest against what he saw as constant excess. In *Walden*, he explores why people worked to obtain more things which required more work. Whilst staying at this house by Lake Walden, he managed to work only one day a week, leaving the rest of his time for reflection and long walks. His habit of walking became an expression of his freedom to use his time as he saw fit. There is no doubt that during these walks he refined his approach to life that helped him to leave a lasting impact.

German philosopher Friedrich Nietzsche was also a compulsive walker. 'All truly great thoughts are conceived by walking' Nietzsche said. He would walk with his notebook almost every day and record the thoughts that came to him. Gros said that:

Nietzsche was a remarkable walker, tireless. He mentioned it all the time. Walking out of doors was as it were the natural element of his oeuvre, the invariable accompaniment to his writing.

It was this act of walking alone, spending time with his thoughts, that allowed Nietzsche to produce his influential works that have stood the test of time. The list of great thinkers who employed walking as a way to process their thoughts is almost endless. To appreciate and experience the benefits of walking that these great thinkers of the past exploited, we need to leave our phones and earbuds at home, disconnect from our screens and experience the pleasure of walking without distractions. This allows the internal conversation to occur that helps us to better understand ourselves and our own views, just like the great thinkers of the past.

Loose Dirt

I want to return one last time to *Heart of Darkness*. We touched on Marlow's encounter with the second-in-command at Central Station, but I purposely left out Marlow's description of the man. For me, this is the best description of a leader who seeks distraction over solitude:

It seemed to me that if I tried I could poke my forefinger through him and would find nothing inside but a little loose dirt.

This describes a leader who doesn't take time for solitude. This is the kind of leader I'm sure you've seen time and time again. This is the kind of leader who somehow occupied the Oval Office for four years. If you really tried, all you would find inside these 'leaders' is a little loose dirt. Nothing really valuable. Nothing original. Nothing creative. Because all they are doing all day long is taking in the noise around them and reacting to whatever piece of information comes their way.

I'm not saying that everyone who operates under the constraints of screens cannot be successful. Consider both characters we've looked at from the novel Heart of Darkness. One is the manager and the other is the second-in-command. They are 'successful'. They are in leadership positions. But that doesn't validate their worth or validate how they got there. Nor does it signify an ability to lead. Leadership is not found in a distracted mind that is consumed by the noise of everything around it. True leadership requires time spent in solitude. But if we want leaders who spend time in solitude, we need to first make the decision to put down our own screens and spend time in solitude in any one of the ways discussed in this chapter. To borrow a phrase from the pandemic, we need to 'self-isolate' so that the virus of distraction that's been infecting our leaders will stop spreading.

Chapter 5

The Loss of Education

Education is the most powerful weapon which you can use to change the world.
Nelson Mandela

For six years of my school life, I was homeschooled. Don't be quick to judge. We weren't your typical off the grid, raise some chickens and make our own clothes homeschoolers. We were just a regular suburban family who, despite living right next door to a school, had decided to homeschool. Admittedly, school at home had its benefits: you never really got after-school detention; you didn't really get homework; and watching television at lunch when you're a twelve-year-old feels like a dream come true. But, by the age of fourteen I had outgrown home school and was forced to start 'real' school. I vividly remember the first few hours of my 'second' first day of school. My first class was with a teacher called Mrs Slaughter, which didn't make for a promising start. She turned out to be a lovely teacher, but I did initially worry for the kid in the class whose last name was Lamb.

Our school and our teachers have a significant impact on our lives. For the majority of our young lives we are defined by school and our connection to it. As we grow older, this connection weakens, but it still inevitably persists. Schools often hold reunions for decades after students graduate. This illustrates the special place we hold in our hearts and minds for our experience of school. When my wife found out she would be teaching kindergarten, she was initially apprehensive, and understandably so. For most people, the thought of managing a group of five-year-olds as a job sounds more like purgatory than a rewarding career. But I tried to highlight how most people remember their kindergarten teacher for the rest of their lives. It has been well over twenty years since I had Mrs Cunningham as a teacher, but she is still firmly fixed in my mind. Unfortunately, when I surveyed a few of my wife's family members about whether they remembered their kindergarten teacher, I received mixed responses, which didn't help my cause. But, to her credit, my wife taught kindergarten for two years before stopping for the birth of our first child. Now back at work, the children she had in those first two classes adore her, constantly coming up to her in the playground and wanting her full attention.

My wife's experience of teaching was a big factor in my attempt to obtain a postgraduate degree as a teacher. Through my studies, I was exposed to the world of education in an eye-opening way. The education system has, like many other things in our world, fallen victim to the promise of a better world through the use of technology and screens. Distraction has become commonplace in our classrooms, and educators across the system, from classroom teachers to government leaders, under the constant pressure of technology companies, have sought only to include more of it. Parents are often given no better alternative and are forced into embracing technology at each stage of their child's educational journey. To do otherwise would mean being labelled out of touch and would risk disadvantaging their child. There are few things that any decent parent would want less than to put their child's education at risk.

The growth of the education technology industry, or EdTech as it is commonly referred to, has been dramatic. And the amount of money up for grabs in the world of education is the first in a long line of issues that should be explored before allowing screen devices to totally take over the education of the next generation.

Transforming Teaching?

Trying to find accurate data for the size of the EdTech industry is like guessing how many lollies are in a jar. Everyone seems to have a different opinion and you never get to find out who is right. Some websites estimate that the value sits around US$227 billion, so we will just go with that. The lack of precision is partly explained by the reality that a lot of forecasting is terribly inaccurate and full of assumptions. But it can also be explained by the differences in what exactly is included in the definition of EdTech. For our purposes, we'll consider EdTech to be any form of technology that permeates into a typical K–12 classroom. This can include anything from access to a computer room at school and having a smart or interactive whiteboard in the classroom, all the way up to every student having a laptop as part of the learning experience in every class.

Regardless of how you define the industry, one thing that all sources agree on is that it is growing at a rapid pace. Estimates put the annual growth rate at about twelve per cent. The investment from venture capital firms is also on the rise, which is notable given the kind of returns these firms try to achieve. In 2014, venture capital firms invested about US$1.8 billion in EdTech companies. The amount peaked in 2021 at a staggering US$20.8bn.

It is important to consider the motive of these venture capital firms. They are notorious for demanding large returns from their investments. Typically they will invest early on in the life of a business in the hope that their investment will produce a substantial return – ten, twenty, even one hundred times their initial investment – as can happen with start-up companies that seek to grow at a rapid pace and have their company valuation inflated to a level that would make the founders and the venture capital firm that has invested in the company very rich. From there, the venture capital firm takes their profit and starts again with another set of companies. The point is that venture capital firms aren't typically in it for the long run. They aren't looking to fulfill the vision of the company. They are looking to invest for a few years and turn a substantial profit, a mentality that seems in conflict with the interests of education. Education shouldn't be about making a profit, let alone making a quick profit. Education is a lifelong process. But venture capital firms and companies alike are taking a profit-focused approach to EdTech.

One particular company worth discussing is Pearson Education. Pearson has long been a player in the education market, known for their publishing of a vast array of textbooks and hard copy educational materials. But in the last decade, Pearson has shifted its focus to the digital education market, in an attempt to pivot to changes in demand. This shift has certainly cost the company in the short term. During 2017, after several blows to the business, including the loss of a number of testing contracts and a downturn in textbook sales, Pearson suffered its worst annual loss (£2.5 billion). But Pearson CEO John Fallon remains hopeful that the short-term pain will be worth it in the long-term. When speaking about the EdTech market, Fallon remarked that 'there is going to be a big winner and we are absolutely determined that Pearson is that winner'.

Pearson has been playing the game of trying to win in EdTech for a long time, declaring to shareholders back in 2013 that 'education will turn out to be the great growth industry of the twenty-first century' and that an investment in education will 'pay the best interest'. Pearson is a company that clearly prioritises shareholder value above all else, just as any self-respecting company in a society built on the principles of capitalism should. But it clearly raises some questions when considering what industry they are playing in. Education should not be focused on building the education systems that are most profitable, it should be focused on enhancing the learning of the student, regardless of what that entails, be it profit or not. But Pearson seems to have a laser focus on implementing what they see as two key changes to the education system, both of which rely completely on technology.

Firstly, they want to disrupt teaching. Disrupting has become a popular term used when up-and-coming businesses attempt to change (disrupt) how an industry, profession or business model operates. It's built on the premise that disruption is good and necessary, that the industry in question needs this disruption because it is failing in some capacity. And what is implied in the notion of disruption is that the proposed change will improve the industry by introducing a new model that takes or improves what is currently good and fixes or removes the problems that currently exist. But that is rarely the case. Often the disruption, in attempting to fix the problems of an industry, introduces its own set of problems, resulting in an industry that in a few years will be primed and ready to again be disrupted.

When we hear how a company is disrupting an industry, it is important that we don't simply embrace the notion that it's needed or that it's good. Michael Barber, Pearson's chief education advisor, and co-author Peter Hill, state in a 2014 report that teaching is an 'imprecise and idiosyncratic process that is too dependent on the personal intuition and competence of individual teachers', and that this problem can be fixed by 'overthrowing' and 'repudiating' the 'classroom teacher as the imparter of knowledge' and replacing them with 'increasing reliance on sophisticated tutor/online instruction'. Pearson is not trying to replace the teacher entirely, rather they are trying to change their role and rebrand them as facilitator.

In some low-fee private schools in Africa, India and parts of South-East Asia that are run by a company supported by Pearson, teachers are required to read prefabricated lessons word-for-word from a tablet device. Staff in these schools cannot deviate from the script and must deliver the learning activities in a step-by-step fashion. In other words, teachers in these schools are not being allowed to think for themselves. They are losing the opportunity to use their knowledge, skills and creativity in the education of the children in their care and must let technology do the thinking for them.

The second change advocated by Pearson is that students be given personalised instruction by the means of artificial intelligence (AI). They have developed a system called Pearson Realize, which they say will help guide and direct student learning in a personalised way. Pearson Realize is built on a database that the AI system draws on to determine what is best to teach each student given their current knowledge and any identifiable gaps. Learning resources are provided to students based on searches for materials within their learning database that attempt to closely match students' learning needs to content, drawing on a vast library of both purpose-built and commercially available materials.

Pearson further argues that the current structure of schools does not meet the needs of parents and guardians. They have created the online Pearson's Connections Academy which claims to offer students the opportunity to reach their highest individual potential by uniquely customising the learning program. The Connections Academy is an online school which offers parents the opportunity to provide their children with an education Pearson says is equivalent to the K-12 education they would receive if they had attended school in person. The Connections Academy offers students the ability to tailor their learning and move at a faster pace if needed, as well as opportunities to meet other students who attend the virtual school through virtual clubs and the occasional field trip.

The Connections Academy places a great deal of emphasis on the parent taking the role of Learning Coach. It's a role similar to the ones fulfilled by the teacher who can only instruct students through prefabricated lessons. Pearson puts the responsibility on the parent to monitor the child's learning activities, test their understanding and make sure they finish the lesson. Effectively, the parent is responsible for ensuring the child is engaging with their screen device appropriately during school time and not playing games. Through Pearson Realize and the Connections Academy, Pearson is not only trying to change the role of the teacher, but remove the need for the teacher altogether.

It's important not to forget their goal, which is fundamentally the same goal as the Venture Capital firms who are pouring money into EdTech. Pearson's goal is to grow shareholder profits by being the winner in the game of EdTech. It is not a goal that aligns with the needs of students and educators. The consequence of Pearson and other players in the EdTeach industry being profit focused will be that students spend even more time trying to learn whilst staring at a screen rather than interacting with teachers and other students in a classroom. But, as we will come to see, interaction within the classroom is how students really evolve into true learners.

Do You Want a Revolution?

In 2007, former Prime Minister of Australia, Kevin Rudd, announced an AUD$2.4 billion program called the Digital Education Revolution or DER. According to the review of the DER conducted in 2013, it had three broad goals:

1. Deliver sustained and meaningful change in the way teaching and learning is conducted in Australian secondary schools, focusing on four strands of change (infrastructure, leadership, teacher capability and learning resources).
2. Provide every student in Years 9–12 with access to technology required for contemporary learning.

3. Create the foundations for effective delivery of an online, nationally consistent curriculum as well as providing stimulating and challenging learning resources for students.

Overall the report summarises that 'the DER was designed to generate the broadest possible impact, and the scale of investment was intended to rapidly level the playing field in effective integration of information and communication technologies (ICT) into teaching and learning.' To achieve this, over 900,000 laptops were delivered to students around Australia for their use in the classroom. No change to the education system is made without the underlying goal of improving the ability for teachers to deliver content and students to learn that content. The DER was a clear attempt at both and overall a successful DER would see teachers and students using technology to achieve better learning outcomes

The revolution ended in 2014 when the government stopped funding laptops for schools. Instead, most state education systems and private schools adopted a 'bring your own device' (BYOD) policy, where students were required to arrive at school with a device in hand capable of being used for screen based learning activities. The BYOD policies essentially allowed the revolution to continue. However, these BYOD policies were much broader than the DER. Students would be either strongly encouraged or required to bring a laptop or iPad to school, often from the first year of high school.

Given many years have passed since the revolution started, it is possible to now pause and reflect on whether it has been successful. A rich source of data when it comes to education in Australia is the National Assessment Program – Literacy and Numeracy (NAPLAN). The NAPLAN tests are taken by Australian school students in Years 3, 5, 7 and 9 to examine their literacy and numeracy. The results are compiled and made available online in the hopes that students, teachers, parents and policy makers can make better decisions about education. When you compare the results from the period 2008 to 2018, the period that directly overlaps with the DER and the broad adoption of BYOD programs, what you see is that high school students, those students who would have benefited most from the initiatives, show no statistically significant improvement in their reading ability. Writing was only assessed from 2011. When the results from that year are compared against the 2018 results, there has actually been a statistically significant decrease in the average achievement level of both Year 7 and Year 9 students. The average achievement in numeracy increased slightly between 2008 and 2018 for Year 9 students, but that's the extent of improvements. Billions of dollars later, and only a slight increase in the ability of Year 9 students in their proficiency with numbers.

Another core goal of the DER was to create the foundations for effective delivery of an 'online, nationally consistent curriculum'. The success of this goal can be examined when you look at how the education system fared when almost every Australian school student was, at some point, forced into an online learning environment during the COVID-19 pandemic. The quick transition from face-to-face to remote learning exposed a variety of difficulties for educators and their students. A study of over 1200 teachers at the time of the pandemic found that poor internet access was one of the key difficulties for teachers with just ten per cent of teachers saying their students had access to reliable internet at all times. The study notes that teachers felt ill-equipped to deliver learning remotely. It cites examples of teachers who would drive around dropping off learning materials to students and expect that they would be returned to the school for marking once completed. This tended to happen in lower socio-economic areas where some students struggled with internet access and access to appropriate devices.

What is perhaps the most disturbing finding from the study is that three out of four teachers were of the view that remote learning had negatively affected students' emotional wellbeing to some degree, reporting instances of anxiety, withdrawal and loss of connection with friends. Even if the revolution had been successful in helping students transition effectively to learning remotely, this is a stark reminder of the limitations of education via a screen. In spite of the government's initiatives over the past decade, schools, teachers and students were not ready for the change to online learning. But it's something they should have been ready for. If schools were good with using technology and the tools it provides to educate children, then the transition to digital learning should have been seamless. If we embarked on the DER more than a decade ago, surely we would have seen its fruits by now.

Another way to see if the DER has produced any fruit is to compare how different countries have integrated technology into their education system and see which students have produced better outcomes. The Programme for International Student Assessment (PISA) is a great source of data for this sort of analysis. This programme assesses the proficiency of students aged fifteen years in the categories of mathematics, science and reading. The test is conducted every three years and aims to help policymakers improve the decisions they make regarding how best to enhance the quality of education for the students in their country. Countries are compared against each other, and so it is possible to see the effectiveness of different approaches to education. The 2021 test was postponed due to the pandemic and the 2022 results are yet to be fully analysed, so the latest set of reporting we have is 2018. What's noticeable about Australia's results is that in relation to the rest of the world, Australia is not improving. Up until 2018, Australia exceeded the average result for countries in the Organisation for Economic Cooperation and Development (OECD) group. These are all democratic countries that are generally considered reasonable comparisons for Australia. In 2018, Australia's results in the PISA assessment for mathematics dropped below the average results of its peers in the OECD group for the first time. Five countries whose mathematics performance was on par with Australia's in their first PISA assessment now outperform Australia. And, out of sixteen countries whose mathematics performance was lower than Australia's in their first PISA assessment, nine now outperform Australia and seven are now on par with Australia.

The 2018 report on Australia's PISA results also notes that reading has been 'steadily declining, from initially high levels, since the country first participated in PISA in 2000'. In fact, across the board, the results in reading, mathematics and science have been steadily declining since 2000, which is in line with the NAPLAN test results. This is clearly not good, and certainly not the intended result of a revolution.

To help achieve the goal of the programme, PISA regularly leverages its vast amounts of data to produce reports that draw conclusions about education. These reports help PISA participants better understand what factors might help improve the results of students and their ability to learn. A report produced by PISA in 2015 focused on the impact of technology in the classroom. Below is a summary of the conclusions:

This analysis shows that the reality in our schools lags considerably behind the promise of technology ... Where computers are used in the classroom, their impact on student performance is mixed at best. Students who use computers moderately at school tend to have somewhat better learning outcomes than students who use computers rarely. But students who use computers very frequently at school do a lot worse in most learning outcomes, even after accounting for social background and student demographics. The results also show no appreciable improvements in student achievement in reading, mathematics or science in the countries that had invested heavily in ICT for education.

PISA is an organisation independent of government influence. It does not exist to turn a profit. PISA isn't pressured by the media, technology companies, teachers' unions or any other organisations. PISA simply collates and analyses data on student achievement and reports back to its members on what the data suggests is the best way forward. The conclusion of PISA is that there should only be limited use of technology in the classroom. Greater use of computers and tablets does not equate with better learning outcomes. In fact, it would likely be better to stay away from these devices altogether.

Improving educational outcomes is not only a focus of governments but also of many large philanthropic organisations. The Bill and Melinda Gates Foundation is the largest philanthropic foundation in the world. Its founders have worked tirelessly for more than twenty years to address a range of issues. During this time they have achieved some wonderful outcomes, particularly the vaccination of millions of young children to protect them from diseases that might have otherwise cut short their lives. One focus of the foundation since its inception in 2000 has been education, putting more than US$8 billion into the foundation's US program, which focuses mainly on K-12 education. Although this investment has resulted in some improvements, such as an increase in high school graduation rates, other metrics that the foundation uses to judge the program's success, such as the percentage of high school students who enrolled in postsecondary institutions, haven't moved. In 2020, Melinda Gates wrote: 'When it comes to US education, though, we're not yet seeing the kind of bottom-line impact we expected.'

When you first visit the foundation's website for its US programs and land on the K-12 section of the site, you are greeted by a graphic of a young child sitting at a coffee table with their laptop in front of them. The young child has their headphones on and the computer screen is clearly showing the child in the middle of a lesson, with the teacher and other students on the screen. On the other side of the coffee table is the child's mother sitting on the lounge with her legs crossed and also staring at a laptop. The picture provides a clear visual indicator to show the way in which the foundation is looking to improve education. A quick look at the main K-12 education grant recipients reinforces this. The website for one recipient, the International Society for Technology in Education (ITSE), states that 'we help educators around the world use technology to solve tough problems'.

Also on the page is a list of the six different areas the foundation provides funding for. The first area listed is 'catalysing innovation', which is described as 'advancing research and development in education technology and learning to accelerate progress for Black and Latino students, and students experiencing poverty'. The foundation's push for equal opportunity across race and socio-economic status is evidenced throughout their program as they always aim to 'lead with equity'. But the foundation sees the solution to the problems outlined in that focus area as coming from research and development in education technology. When you combine the image, the grant recipients and the program's first focus area, you see that the foundation is looking to lead not just with equity, but with technology.

There is a lot to admire about what the foundation is doing. The foundation is empowering teachers and principals and trying to do so not for the purpose of profit, but for the purpose of improving student education. But one of the main ways they are looking to improve student education is through the use of technology. Again we see the underlying assumption that technology is the solution to the problem. The foundation is embracing the mindset that all of life's current problems can be solved through smarter technology. This response is understandable. Technology is what Bill and Melinda Gates are good at. It's what built their wealth in the first place. But as we've seen, investment in technology in order to improve the educational outcome of students is an investment that yields very little return for students themselves.

Technology in the classroom can take many shapes and forms, but predominantly it will involve the use of a screen to communicate information to students. The medium is often equally as important as the message. The nature of screen devices reveals why learning with them is so difficult, a point that is perhaps being missed by the foundation and many education policy decision makers. Screen devices provide endless opportunities for distraction but learning requires periods of sustained focus. It requires the student to contemplate how a piece of information fits in with their broader world. Trying to do that on a device that is designed to distract is a big ask. The call of distraction is too strong for most adults to resist, let alone school children. When children get distracted, they stop learning and their education suffers. But the system expects students to self-regulate, to learn how to manage their learning whilst having a screen in front of them.

One Sydney private school that enforces a BYOD policy for all students from the age of ten years says that it is important for students to learn to self-regulate and to develop self-control in avoiding these distractions. It is quite likely the adult who wrote that quickly turned to answering an email or replying to a text message before moving onto the next sentence. You don't learn to self-regulate when on a device simply by using it. That would be like trying to teach adolescents to regulate their intake of alcohol by giving them an endless supply. The addiction that is distraction requires that we learn to value focused attention and then implement ways to mitigate distractions when we are using devices designed to distract us. Children will never learn these things unless they are explicitly taught them and are helped to safely navigate the situation. This is because, unlike other dopamine inducing activities, screen addiction has no feedback mechanism to help with self-regulation.

We explored the biological feedback mechanism involved in other dopamine inducing activities, such as eating chocolate, in chapter two. We don't get the same obvious biological response we receive from other addictive behaviours as we do when we overuse screens. This is a large part of the reason why managing addiction to screens is so hard. Consider this in the context of evolution. It is estimated that it takes around 50,000 years for us to evolve any feature that helps to deal with a change in our environment. The internet has only been around for about twenty years, television for about sixty years. So we've got a few years yet to go before we evolve a feedback mechanism that helps us to deal with these dopamine producing activities. From a biological standpoint, no human is in a position to learn how to deal with this constant dopamine exposure.

It is even more difficult for children than it is for adults. Children's brains are not as developed as adult brains. What is particularly underdeveloped is the part of the brain that is heavily involved in self-regulation and self-control, a region of the brain known as the frontal lobe. This region isn't fully developed until a person is well into their twenties. A child simply does not have the neurobiological infrastructure in place to handle the level of self-control and self-discipline needed to effectively learn in an environment where screens are present.

The solution to a better education for our children does not lie in technology. The proliferation of screens in the classroom has only served as a distraction from where the real focus should be for philanthropic organisations and government officials. Better teaching starts with better teachers.

Educating Educators

At the heart of teaching is the teacher. The Bill and Melinda Gates Foundation touches on this in seeking to provide teachers with 'in-school learning and leadership opportunities'. But this line is tagged on at the end of a technology-centred mission. Teachers should be the first focus of policy makers and philanthropic foundations, as the best way to improve student learning is to provide them with good teachers. I'm not just saying this because my wife is a teacher or that I spent time training to be a teacher. It is because from the very start of the journey to becoming a teacher, you are exposed to research that clearly demonstrates teachers are the most important external factor in a child's education.

John Hattie synthesised over 800 meta-analyses of quantitative data relating to student achievement. His synthesis covered studies involving millions of children over a span of fifteen years. The size and breadth of this study is enormous, giving significant credibility to the results and conclusions because, as with any study, the larger the sample size, the more you can infer that the results are an accurate reflection of the population.

Given Hattie's study was a synthesis of research involving millions of students and their interactions with their teacher, his conclusions can be considered to have substantial value. What he found to be the biggest factor to influence student learning is the students themselves. The character and willingness of students accounts for fifty per cent of the variance in student results. But coming in a close second is the teacher, who accounts for thirty per cent of the variance in student performance. Hattie identifies other factors, such as peer interaction and the school itself, but nothing comes close to the influence of the teacher.

There are several other studies that demonstrate the difference a good teacher can make on student learning. Economist Erik Hanushek created a model that showed that if a student has a good teacher for five years as opposed to an average teacher, the effect would be enough to close the gap associated with low socioeconomic status. Whilst quality teaching can help to remove the disadvantage created by low socioeconomic status, technology in education often emphasises it, as we saw when learning went completely online. Another study, this time undertaken by Australian researchers, showed that students with effective teachers can learn in three-quarters of a year what would take a less-effective teacher a whole year to teach. The effect of quality teaching is also lasting. Achievement gains have been found to last for three years after exposure to an effective teacher.

The evidence in favour of quality teaching is overwhelming. If you want students to learn better, help teachers become better teachers. Invest money and time into teacher professional development. Improve teacher pay and conditions to incentivise more individuals to become a teacher. Help them understand their topic area better. Keep them passionate about teaching. Help them learn new ways to engage students and combat behavioural management issues in the classroom. Don't just leave them to fend for themselves. But the focus of those responsible for the education system still appears to be on integrating technology into the classroom. And the messaging is that if you are not in favour of technology in the classroom then you must be a 'technocynic', a label I was given in one of my subjects at university after responding to a survey question. 'On a scale of one to five, how willing are you to integrate technology into the classroom?' I happily selected 'one'. I was the only student in the class who did so. But other technocynics exist. In fact, they are often the very people selling technology to educational institutions.

Technocynic Central

In chapter one, we saw how some of the biggest names in technology put strict limits on the technology their children can use at home. When it comes to education, we find a very similar story. A lot of technology professionals in Silicon Valley are opting to send their children to schools with little or no use of technology in the learning experience. An example of one of these schools is the Waldorf School of the Peninsula, where students learn without technology or computers in the classroom. The teaching philosophy is instead focused on physical activity and learning through hands-on tasks.

In the early childhood programs, Waldorf schools focus on practical activities that foster creative thinking and imagination. In the middle years of school, there is a strong focus on the creative arts, with students participating in art, drama and music. They also place a focus on building the social capabilities of their students. The focus for high school students is the development of critical thinking and independent judgement.

Seventy-five per cent of students who attend the Waldorf School of the Peninsula have parents who work at a technology company, such as Apple, Google, Facebook or Microsoft. It's saying a lot that parents who are exposed to technology as a part of their everyday life would look to provide their children with an education that focuses on developing and learning without it. What is it saying about the technology they are developing? What does it say about what they really value for their children? Not that they spend each day learning via a screen, but that they have teachers who will help them to be creative and think for themselves.

The technocynical can also be found in one of Australia's best schools, Sydney Grammar School, which is consistently ranked in the top ten schools in the country in terms of academic performance. Sydney Grammar School does not expose its students to any technology in the classroom at all. Phones, laptops and iPads, are completely banned from the school. In fact, up until Year 10, students are required to hand-write all their assignments. The former headmaster of the school, Dr John Vallance, was quoted in a newspaper article saying that the billions of dollars that the government spent on the Digital Education Revolution was a 'scandalous waste of money'. The current headmaster, Dr Richard Malpass, holds similar views about technology in education.

Sustained student attention is essential ... The modern device, notwithstanding its almost incalculable internet reach, is nonetheless the triumph of potentially near-infinite distraction with the temptations of social media and the sheer multiplicity of information on offer instantaneously and ever-temptingly in quick view.

For Dr Malpass the heart of effective learning is effective teaching.

As an observer of so very many lessons delivered by other teachers, it has become abundantly clear that the most effective lessons were those in which the teacher's presence, his or her subject knowledge, that palpable passion and communicative warmth in sharing the experience of their topic simply inspired the room.

Good teaching doesn't mean pointing students to a website or a google document; it means inspiring the room!

We yet again find ourselves navigating our way back to the conclusion that technology is taking some of the most important things in life from us by promising us a better world, but delivering one that was worse than what it was to begin with. In the pursuit of profits, we are leaving our children's education to devices designed to distract. Instead of filling our classrooms with technology, we should be building learning environments that help students to focus and think, and we should be supporting teachers in their profession. A digital education revolution that would actually work would be one that removed the word digital altogether.

Chapter 6

The Loss of Memory

*Remembering can only happen if you decide to
take notice.*
Joshua Foer

We all come to the realisation at some point in our lives that we have been 'born and raised'. Often we will associate the phrase with a specific location, such as 'I'm a born and raised Australian' or 'I'm a born and raised city boy'. But there comes a point in our lives where we are no longer in the process of being 'born and raised' but have transitioned into that state. That point typically occurs somewhere in our mid-twenties, when we can no longer use that common phrase 'when I grow up, I want to ...' (you can fill in the blank).

I am well past dreaming of what I'll be when I grow up. But when I was younger, of all the things that went through my mind, a hand surgeon was the thing that kept my attention for the longest time. Like the majority of people, I didn't end up fulfilling my dream. Only a lucky few manage that. But it's generally not because we fail in some way, but because circumstances change. So it's interesting that we ask children to tell us what they want to be when they grow up. It's as if we believe they can truly position themselves in the midst of a professional life – perhaps as a midwife delivering babies, or as a lawyer defending a case that could see an innocent woman go to jail, or even as a politician representing their country on the global stage. Not only is it a poorly thought-out question because there's no way a five-year-old can really understand what it means to be in any of those professions, but it teaches kids that life is a destination and not a journey. Although the destination determines the journey, life is still simply one day at a time.

My desire to be a hand surgeon has its origins in my time as a homeschooler. My siblings and I often became bored because we were stuck at home. To alleviate our boredom, my mother invented all sorts of games. Most were lots of fun, but some lacked her creative flair. One game, which we only played once because of its consequences, was a race to see who could count all the windows in the house the fastest. It was just my sister and

I playing this game, and losing to her was unthinkable. Luckily, she got to go first, because once I knew her time I had a goal to work towards and was determined to not fall short.

Off went the timer and I raced around the house counting every window. To win the game you had to touch every window in the house. By about halfway, I knew I was making good time and was well on the way to beating my sister. But I didn't want to just beat her, I wanted to thrash her. The final window happened to be in my bedroom. I raced in and in the exuberance of knowing I was victorious over my sister, I hit the window hard. The glass, which was very old, splintered as my hand went through. I automatically pulled it back to find an 8 cm cut going up the underside of my arm from my wrist. The image of the cut, which went right through to the tendon and almost severed it, is seared into my mind. I would love to tell you that I was very brave and calmly told my mother I had hurt myself and we needed to call an ambulance, but the sight of the gaping wound sent me screaming to her hysterically.

I ended up needing surgery to repair the damage I'd done to both the tendon and some nerves, but within three months my wrist was back to normal. My mother befriended the surgeon who operated on my hand. A few years later, when I was looking for a work experience placement, she contacted him. After my injury, I'd set my sights on becoming a hand surgeon, and somehow my mother convinced this doctor to let me spend a week at the Sydney Hospital Hand Unit following him and other doctors around as they consulted patients and performed operations.

It was one of the most incredible weeks of my life, and to this day I can give you a day-by-day run-down of what I experienced – who I met, what I was taught and what kinds of operations I witnessed in the operating theatre. I remember I had to get a photo taken for my hospital pass, and it was the most horrible photo I've ever had taken! I remember that I listened to John Mayer's song 'Clarity' every morning as I arrived at the hospital because the song, especially the trumpet solo, made me feel like a doctor. I can even tell you that for most of the week I ordered a meatball sandwich for lunch, which I can still taste.

All the doctors in the unit told me to spend as much time as I could with the professor as he had been a hand surgeon for decades and would teach me priceless lessons. I managed to spend Thursday morning with him, and he taught me that being a doctor was all about the patient. You had to put yourself into the shoes of every patient you saw and understand how their injury affected their life. Only after doing that could you see how best to solve their problem. According to the professor, being a doctor was never about trying to solve the problem in the best way you thought possible, but in the way that best suited the patient.

My memories of that week are incredibly vivid. But ask me to tell you what I did last week, or the week before that, or the second week in February this year, forget about it. I could tell you general things, like I went to work and I had a coffee or two. But I'd be hard pressed to find a week in my life that I remember in as much detail as the week I spent at the Sydney Hospital Hand Unit.

As a culture, we appear obsessed with remembering every detail of our lives. It's like we are terrified of forgetting. Although we don't talk about this obsession, we display it in our actions almost every day. A great example can be found if you go to a concert. You will almost certainly see people around you who think it is better to spend the whole evening watching the concert through their smartphone screen so they can record the event, than to enjoy the whole experience as it happens. They do this to achieve two things. First, they want to share their evening with all their friends on social media. Second, they want to be able to look at the video and remember what a great night they had. But in the process of taking a permanent record of the evening, they don't really experience it themselves. It's impossible to be fully present at a concert, taking in every guitar solo, every leaping high note and every goosebump inducing crescendo, if you are busy looking for the perfect filter.

Everywhere you turn you see people using their phones to try to capture a moment by taking a photo or a video. We then share this content via social media. Our social media accounts then act as a repository for our memories. We scroll back through them and recall all the fun things we did with all our friends. People build a social media memory bank without even realising it. Fundamentally, all these photos and videos are an attempt to remember the moment, to hold onto something we feel has value. But the insertion of technology into these moments not only ruins the moment itself, but also our memory of it.

How Memory Works

To understand how technology is ruining memory, we first need to establish how memory works. To do that, we will start by taking a step back and considering the journey of memory across the ages.

Our memory hasn't always served the purposes it does today. In fact, memory has served many different roles across time. But each of these roles impinges upon our use of memory today. When we were primarily a hunter-gatherer species, our main goal was to find food. Related to that was the need to get the food back home.

Therefore, one of the most important things for us to remember was our way home – our way back to the tribe or back to the cave. Crucial to long-term survival of the tribe was the need to keep those members who were unable to hunt – babies, young children, pregnant and nursing women – protected and fed. So, the success of hunting and gathering trips depended on the ability to not only find food, but to also get the food back to those who needed it. Evolution favoured those who could both track down the food and remember the way back home.

Despite no longer needing to go out hunting and gathering, we have retained this ability to remember locations remarkably well. Take a moment to recall your childhood home. It may have been decades since you lived there. But if you take a walk down memory lane you'll probably be able to describe in detail almost every aspect of your childhood home. You will be able to tell me what the entrance was like, what the first room in the house was used for, how the living room was organised, where the bedrooms were located, and how the kitchen was set out. If you are good at drawing, you would be able to map out where everything is in the home. This is because, thanks to evolution, the ability to remember locations is built deep within our brains.

Many people use this function of memory to their advantage when they build a memory palace. The best way to explain what a memory palace is and how it is used is to work through an example. Try to remember these five words in order: person, woman, man, camera, TV. These are the same five words that Donald Trump bragged about being able to remember in the cognitive fitness test he took in 2020. Trump claimed that he could recall that list of words in order twenty minutes after he was given it. You might be a little nervous now, thinking if Trump can remember the list then surely you must be able to. To save you going back, here is the list again: person, woman, man, camera, TV. Now you might read that list five or six times, maybe more, and you'll feel like you can close your eyes and recite the list without even looking. The difficulty now is remembering it twenty minutes from now. Once your mind has wandered on to some other topic or has been distracted by something else, what is the chance you'll be able to recite the list in the correct order? Well, whatever probability you assign to that answer, it's about to get close to 100%, because we are going to put that list of words into a memory palace.

Think back to your family home, the one you grew up in when you were a child. Or if you prefer, you can think of your current home. The goal is to think of a location you are very familiar with. This is the key to the creation of a memory palace, because to create a memory palace, you need to be able to visualise a location and place objects within that location that will remind you of the things you will need to recall.

To start building a memory palace to help you memorise the list, picture yourself walking up to the house and standing a few feet back from the front door. Now, at the front door of your home, I want you to place a giant pearl. But picture this pearl like a person. The pearl has a face, maybe even a moustache, and it's standing there, staring at you, on two pearl legs, with two pearl arms and a big pearl smile. Now that pearl is accompanied by his son. You know it's the pearl's son because he's got one of those little helicopter hats on, the colourful ones with the propellers, and the propeller is spinning really fast. The 'pearl-son' is also carrying a tiny little backpack as if he has just walked out of the house through the front door and he's on his way to school. You greet the pearls, and you move inside.

Your next step takes you to the dining room. It looks odd, however, because there is a woman sitting in a yoga pose on the dining room table. This woman can be whoever you like – a crush, a celebrity – but you need to think of someone who stands out in your mind, someone you can easily picture. You notice this woman, not only because it is unusual for someone to be on your dining room table, but also because she is doing yoga. You stand in the dining room for a moment staring at her, and she stares back. And every few moments she changes her yoga pose.

Next you turn to the couch and notice a group of men sitting on it. This is not just any group of men, it's three of the most famous men in the world: George Clooney, Brad Pitt and Matt Damon, each wearing a well-cut suit. Now, your two-seater couch isn't really big enough for all three to sit together comfortably, so they are jammed together almost sitting on top of each other. And each one is constantly struggling to maintain their position on the couch.

From there you head to the kitchen, where you find a photo shoot is going on. One photographer has opened your fridge and is closely inspecting each item in the fridge with a camera that has an outrageously long lens, a lens designed to shoot pictures of objects hundreds of metres away. But there the photographer is using it to take a close-up of your tomatoes. You try to get the photographer's attention but their face is buried in the camera, which they are holding with two hands. They are certainly too busy to acknowledge you.

Finally, you get tired of this madness, and you head off to your bedroom. You are greeted by a television sitting on your bed. But it's not just any television, it's the purple television from *The Simpsons*. It's got those funny looking bent antennas and antiquated dials down one side. On the screen of the television is the Simpsons family themselves, and they are waving right back at you.

You've just created your first memory palace, a place you have created in your mind using a location you are familiar with and filled with images you associate with the words you need to remember. I have also given you a few extra tricks to help you remember the words. When you picture things in a memory palace that are moving, you tend to remember them better. The pearl-son's hat has a spinning helicopter rotor on it, the woman is changing yoga poses and the three Hollywood hunks are moving around on the couch. The woman and the couch full of Hollywood hunks are examples of another trick that can be used in memory palaces, which is to choose items with sexual or sensual overtones to help you remember. All the items in the memory palace were unusual in some way, to help them stand out in your mind, but choosing something familiar to put into your memory palace, such as the television from *The Simpsons*, can also help you to remember the item.

So now that we've set-up your memory palace, go for a walk through it. Greet the pearls at the door, take a moment to watch the woman doing yoga on your dining room table, greet Hollywood's finest men, try to get the attention of the cameraman in your kitchen as he takes a close up of your tomatoes and enjoy *The Simpsons* in magnificent high definition on the purple television that is conveniently in the middle of your bed.

If you do this a few times, you won't only remember the list in the correct order twenty minutes from now, you'll remember it tomorrow, in three days' time and probably into next week. At first memory palaces are difficult, they require you to focus and be creative. But once you get better at them, you begin to be able to remember things in incredibly short periods of time. At the world memory championships the best mnemonists in the world can memorise the order of a deck of cards in under twenty seconds using memory palaces. One can only imagine the level of concentration and focus required to get that good at remembering.

Despite this profound ability to remember, a modern narrative is often that we should distrust memory, that it isn't reliable. Often, when the ability to remember accurately is challenged, the example of eyewitnesses to a crime gets brought up. If a group of people witnessing a crime remember the event in vastly different ways, how can we trust ourselves to remember anything at all? This approach to memory is too narrow. Yes, our memory can fail us, and it often does. But the exercise we just went through shows how powerful our memory can be.

For the majority of human history, memory was crucial for storing information and passing it from person to person and generation to generation. This was most often in the form of stories. These stories were vital for teaching culture and tradition to the next generation.

The stories that were memorised and passed on orally were the building blocks of ancient societies. They created a shared understanding amongst members of the group. And that storytelling relied on memory.

Memory has long been relied on for recording, recalling and retelling the details of events and for passing on knowledge. This oral tradition, which relied on memory, continued even after the development of the written word, as books and manuscripts were only accessible to the privileged classes. This only started to change with the invention of the printing press in 1440, which helped to make the written records of events available to all.

The invention of writing as a way of recording events and information raised an issue that hadn't previously existed. What is the effect on memory? Famously, Socrates in his discussion with his friend Phaedrus warned of the dangers of relying on writing things down:

They will cease to exercise memory because they rely on that which is written, calling things to remembrance no longer from within themselves, but by means of external marks.

Clearly our screen devices are not the first invention to change the way we use our memory. Nor are they the first invention to take something from us. This may sound dramatic, but it demonstrates the value people had for their memory prior to the introduction of the written word. What was written down was no longer theirs to keep. It was only given to them by 'external marks'.

In an eerily similar way, our memories are no longer ours. They are now given to us by external marks. Be it a picture, a video or a written post, our memories are offered up for the approval of the masses, and their value is determined by the reception they receive. Things that were once retained in memory are placed online, not so much to help us remember who we are and where we came from, but that we might have value attached to them because of the reception they receive from others. Our memories only seem to matter now if they are liked, loved, or commented on.

Our New Memory

To truly understand how technology has replaced and changed our memory, it's necessary to understand how human beings use tools. I'm not just talking about tools used in practical trades such as construction, but tools of all sorts. A musical instrument is a tool. A surfboard is a tool. A chef's knife is a tool. Tools are anything we use as human beings that allow us to do things that are otherwise more difficult or impossible to achieve. Consider again our evolutionary roots, and how the ability to use tools would have been prized. It was the human who could use tools to start a fire, hunt for food or build a shelter who was more highly valued than others because these tools helped with survival. We have become very good at using tools because they help us survive. In fact, we are so well adapted for using tools that often we feel as if they are part of who we are.

Consider an expert musician, someone who has spent well over 10,000 hours honing their skill. When they play, it's as if the instrument is an extension of their body. It's as if they're in perfect harmony with the instrument. When the individual isn't holding the instrument, they look and feel a little uncomfortable. There's something missing when they are without their instrument. And when they pick up their instruments again, their neurons start firing in such a way that their instrument is almost like another limb. Because that's what their brain actually thinks it is.

Don't just think of musicians; think of any athlete, artist or professional who uses a tool, and see how it becomes 'one' with them. The screens in our lives are no exception. Your smartphone, for example, is probably constantly within your reach. If it's not in your hand, it's in your pocket, or your bag, or on the table in front of you. If you can't immediately put your hand on it you go into a mild (or major) panic. Without it you feel totally lost. It's so integrated into your life it has, according to your brain, become a part of you.

When we use a tool, that tool often replaces a function that we used to do ourselves. The area of the brain that controls that function understandably needs to do less work. Why would the brain expend energy doing something the tool is doing? Survival means conserving energy, and our brains are designed with survival as the number one priority. So in order to survive the best way we can, when a tool replaces a function we used to do ourselves, part of the brain does less work and ultimately becomes redundant.

Enter our memory. Why would we bother using any of our brain power remembering anything when we can externalise all of the things we need to remember? Why would we bother trying to remember a location when a picture of that location can take us back there over and over again? We wouldn't. We don't. The use of navigation apps is a poignant example. Rather than looking on a map and memorising a route, we now plug our destination into Google Maps and allow it to do the work for us. And rather than taking note of everything around us so that we will remember the route for the future, we are focused on the voice telling us when we next need to turn or what lane we need to be in. The spatial function of the brain, the area associated with remembering places and spaces and the relationships between them, is being used less and less as we rely on navigation apps.

But that does not mean the capacity to use our memory is not still within reach.

A great example of this can be found when you look at London cabbies. Despite the advances in GPS and navigation technology, London cabbies must still pass a detailed exam prior to receiving their accreditation from the London Public Carriage Office. The exam, known as 'The Knowledge', requires these cabbies to memorise every street and landmark in London. They must then use that knowledge to chart the quickest route between two destinations, naming every landmark along the way. It's notoriously difficult. Applicants can spend years studying for it, and only about thirty per cent of those who train for the exam end up passing.

Eleanor Maguire, a neuroscientist at University College London, wanted to see if this wealth of knowledge had any effect on the brains of the cabbies. In 2000, she performed a study where she took an MRI scan of the brain of sixteen taxi drivers. She found that the right posterior hippocampus, a part of the brain known to be involved in spatial navigation, was significantly larger in cabbies than in an average person. It's a wonderful example of the body adapting to the lack of tools available to it. These cabbies have forced themselves to adapt to the stimuli in their environment. Their body has done this by literally changing the shape of part of their brains. This is a fantastic example of neuroplasticity. Our brains still have this capacity within us if we decide to put down our 'tool'. But technology has started to mould our brains in such a way that they are starting to find no use for this part of our memory because we now have a tool that can do that work for us. Despite our ability to reignite those unused regions of the brain, we simply do not use them anymore. We save time and conserve energy instead of doing that which has helped to shape humanity – remembering.

More Than Just a Memory

Our memories are a fundamental part of who we are as humans, and we can't simply let technology take that away. What's often missed in consideration of our memory is that it is more than just the process of storing and recalling things.

Memory is a tool that helps you stay present, to stay in the moment. When you commit things to memory you are focused and present, and therefore your experience of the world is all the more vivid, vibrant and engaging. You cannot deliberately remember something without forcing yourself to be in the moment. We often talk about 'living in the moment', and it's plastered across Instagram and Facebook posts every day of the week. But you cannot live in the moment whilst you are living on a screen. When you live via your screen devices, you are telling your brain it can switch off part of its memory function. And when you externalise your memory function, recording, storing and disseminating your memories using digital tools, those memories are transformed into something which holds value only in the approval given to them by other people.

Imagine how you would feel in the following situation. Let's say you posted a photo of your family celebrating a birthday. You've had a great dinner at a fancy restaurant and you want to share the experience with your social media network. So you post the photo, but no one interacts with it. No likes, no comments, nothing. Now you know you've achieved what you thought was your original goal - sharing your memory with your closest friends - because people have liked your photos before and you just intuitively know that people check their social media, so they've definitely seen your post. But it goes unrecognised. This starts to play on your mind. Why did no one like the photo? Do I look funny in the photo? Is there something wrong with me? Is there something wrong with how my family looks? You wonder why it did not get the appropriate reaction from your friends and this takes away from the memory and experience itself - a nice dinner celebrating a loved one's birthday. You could argue that no photo will ever go unrecognised, but this hypothetical example shows that the act of posting is not just about sharing the memory, it's about receiving approval for it as well.

Being present is where life is best lived. When I'm with my daughter, I don't want to be anywhere else. I don't want to be thinking about anything else. I want to be right there, enjoying everything she does in every stage of her life, fully embracing and taking in every smile and laugh, holding onto every cuddle and kiss. If I hold a camera up to my face as she cuddles me and then post the photo on Instagram with a 'my baby girl' caption, it means I've missed the point of that moment entirely. I don't see any value in taking that photo and posting it, for her or for me.

I also know that my daughter is fully present at that moment. She is a toddler, so there isn't much on her mind. When she sees me after I've been at work for the whole day, she runs up to me and gives me a hug and shouts 'dad'. For me to be thinking about how the moment would make a cute Instagram post completely eradicates the value of it. Because if that's what I am thinking, then I'm not present. And if I'm not present, then where exactly am I?

Our screens want to take us away from the present. They want us to focus on other people's memories, and they want us to focus on how we can gain approval for our memories, for what was once our present. In essence, we are asking for other people's approval of our present. And by asking for other people's approval of our present, we are asking for their approval for our existence.

Our memory is our tool to fight back. By using it we reclaim our past and start to own our present. Your life is the present moment, and it's all you have. Use your memory to fight for your present. As C.S. Lewis said:

Of the present moment, and of it only, humans have an experience analogous to the experience which [God] has of reality as a whole; in it alone freedom and actuality are offered them.

Chapter 7

The Loss of Leisure Time

Lightweight baby!
Ronnie Coleman

For too long I have had a very unhealthy obsession with lifting weights. At the age of thirteen I bought a bench and a barbell and started dreaming of looking like Arnold Schwarzenegger. Alas, due to a combination of the wrong genetic makeup and an enthusiasm for computer games, I never quite made it to the Mr Olympia stage. In more recent years, I have spent too much time learning the style of lifts used in Olympic Games competitions. These lifts are called the 'snatch', where the loaded barbell is lifted overhead in one fluid movement, and the 'clean and jerk', where two separate movements are used to lift the loaded barbell from the floor to over the head.

To be good at Olympic weightlifting, it helps if you are naturally strong and a bit on the short side. Arguably one of the greatest weightlifters of all time, three-time Olympic gold medallist Naim Süleymanoğlu, for example, was only 147 cm in height, had very short legs and arms but incredible natural strength. This allowed him to be the only person to ever snatch 2.5 times his bodyweight. Unfortunately, I am neither short nor naturally strong. I'm 183 cm in height and for most of my life have been described as 'scrawny'. And because my limbs are quite long, it makes the whole spectacle of lifting a weight over my head appear very uncoordinated.

Whenever I showed my wife a video of one of my lifts, she would say I didn't quite look the same as the others in the gym. One of my coaches once told me I was getting better, which I thought was a compliment, but then used the fact that I was looking less 'noodly' as the reference point for my improvement. What does it mean to be noodly? Noodly: to have limbs that resemble noodles. Being told I resemble a carbohydrate certainly made me question why I was spending several hours each week in the gym making a spectacle of myself.

Olympic weightlifting is difficult and painful. For days after a session, I would often walk around as if I needed a hip replacement. And muscle soreness that used to last one or two days when I was younger, now lasts three or four. But, in spite of my complete lack of natural talent, weekly pain and, more recently, torn cartilage in my right knee, I'm still working on lifting that barbell off the floor and placing it over my head.

On numerous occasions I've questioned my motivation. And after reflecting on why I continue to go to the gym and try to master Olympic weightlifting, I began to realise there is one aspect of the hobby that calls me back time and time again. It forces me to be present. When I attempt a big lift, I'm not thinking about what email I've forgotten to reply to or who I need to text back, I'm thinking about lifting the weight. The real beauty comes in the few seconds between when I start the lift and when I finish it. For that brief moment, I'm actually thinking about nothing. As the bar moves off the ground and up over my head, nothing else is going through my mind. There are few other experiences in my life that force me to be so present. That's why I continue to go back, and although they might not recognise it, I suspect it is what draws many others to exercise in general. When you are in the middle of a tough session – a long run, or a big lift – you're not thinking about anything else except what you are doing then and there.

Being present is a benefit of many leisure activities. And it's this ability of these activities to force you into the present moment that is part of their appeal. But such activities are becoming more difficult to fully engage in as screens invade our lives. One of the gyms I attended had the television on playing YouTube videos while at the same time the radio blasted out music. The overlapping sound results in noise. You couldn't follow what's on the radio, and you couldn't follow what's on the television. You can hear both, which also means you can hear neither. If you attend a typical commercial gym, there are no doubt countless television screens around you constantly trying to grab your attention. Many people whilst at the gym are scrolling through their phones or taking videos of exercises to post to social media. Going to the gym and exercising is just one example. Many activities that were once opportunities for us to disconnect and recharge are being twisted into something that is devoid of their original value.

This Is a Call

Dave Grohl is arguably the most influential musician of the last thirty years. He is in every sense of the word a rockstar and is probably the biggest rockstar in the world today. If you listen to rock n' roll music of any sort, you are more than likely familiar with his band Foo Fighters and the band that serves as his original claim to fame, Nirvana. He has also had several music projects on the side, including other bands such as Them Crooked Vultures. In all he has recorded sixteen studio albums, eight compilation albums and collaborated on over twenty-six others. He is often referred to as the 'nicest man in rock', but I think a better description might be the 'hardest working man in rock'.

Grohl attended his first rock concert when he was just twelve years old and was immediately hooked. He spent his teenage years teaming up with friends to form various different bands. During these years he fostered a passion for music that would form the basis of his existence. In a biography written by Paul Brannigan, Grohl speaks in glowing terms about some of the bands that influenced him as a young musician. Grohl describes the moment when the B-52s played 'Rock Lobster' on *Saturday Night Live* as a moment that changed his life. He was so engaged with the performance that he recalled it decades later and spoke to it as a key moment in his musical upbringing. Brannigan mentions a story of Grohl who, while attending a lacrosse camp at the University of Maryland, meticulously saved up money to purchase an album by the band Angry Samoans called *Back from Samoa*. Grohl couldn't wait to get home to listen to the album. When he did, he listened to it over and over again. Even when he was meant to be giving his attention to something else, Grohl had his attention on music.

Another formative band for Grohl was the hardcore punk band Scream. Grohl became the drummer of Scream when he was only seventeen. Grohl described his first audition with the band as 'heaven'. He went on to describe Scream guitarist Franz Stahl, the only band member present at the two-hour long audition, as one of his 'heroes'. Grohl speaks in similar terms about Bob Mould, the lead singer of another influential band in his life called Hüsker Dü. Grohl describes Bob Mould as 'a legend ... an American hero'. Music was and is something Grohl treasures. The performers he listened to time and time again hold a special place in his heart. Most of them are completely unknown to the broader public. But the fact that legendary rockstar Dave Grohl appreciated these musicians gives them a value that would not have otherwise been there.

It's hard to take in just how much Grohl loves music. That's because the way he experiences music and the way most of us experience music are two entirely different things. Throughout Grohl's life, music has been the only thing he has focused on. That's the way he has always known music. But music for most of us is something entirely different. Music is the addition. It's the add-on to what we are doing. It accompanies us in the car. It fuels our intense workouts. It provides the ambience for our many scrolling sessions. It is the background to our conversations at bars. We listen to it as we work. And it stirs our emotions during movies. These days music is seldom experienced simply by itself. With the advent of subscription services such as Spotify and Apple Music, bands that changed the life of one of the most influential musicians of the twenty-first century now simply fill the background of our lives with noise.

Consider Grohl's experience of saving up money to buy a record, of waiting to get home to listen to it and then spending the weekend playing it over and over again – that record and only that record. Grohl gave that record his full attention, but we now rarely give our full attention to music. In fact, we barely pay any attention to it. We add it to the distractions that surround us. The screens in our lives manipulate music into an 'addition'. In doing so, the real value of the music evaporates.

I've played guitar for probably twenty years. That sounds like a long period of time, and you might be fooled into thinking that I know my way around a fretboard. I assure you, that's not the case. But I've been around other musicians who are very good at playing their instruments. These are individuals who have well and truly obtained the 10,000 hours of practise required to master their respective tool. The level of detail and attention that goes into the sound they produce is incredible. Guitarists analyse how the type of wood used in the construction of a guitar affects the sound it produces. Then they look at different types of pedals which you can plug your guitar into to produce different sound effects – distortions, delays, reverbs. All this occurs before the sound is passed through an amplifier – yet another opportunity for musicians to obsess over dials, tubes, and turning the volume up to eleven.

Drummers face a similar plight in their search for the exact sound they are after. Every single aspect of the drum kit is scrutinised in order to deliver the highest quality tone for listeners. From which particular instrument in the kit to use – cymbal, snares, floor tom– to more intricate details, like how tight to have the skin on the drum, where exactly to strike the drumhead and what to hit it with. Many professional drummers even take into consideration the type of sound the room produces when recording an album.

The Foo Fighters produced a documentary series called *Sonic Highways,* where they travelled across America to famous recording studios, re-telling the story of each of these places. This series was also accompanied by a six-track album the Foo Fighters produced, with each track having been recorded in a different studio they visited. The premise was that each studio had its own ambience, a unique way in which the sound was captured, and this uniqueness ended up in the song that was produced. The song could only sound the way it did because it was recorded in that specific music studio.

I understand that such intricate details are hard to appreciate, particularly if you aren't familiar with even the basics of music. This also applies to other activities.

It's hard to watch a sports game when you don't know the rules. And it's hard to appreciate art hanging on the wall when you know little about techniques, the artist or the history behind the artwork. But music is something that is a part of our lives, and it is accessible in ways that most things are not. We don't get into the car and bring up a painting by Michelangelo and stare at it as we drive. We instead save up thousands of dollars to travel overseas to admire his talent as an artist. We don't go to a drive-thru to grab a six-course degustation meal with paired wines. We sit at restaurants for several hours to enjoy the full experience. The value in these things is found when we just do them. Music is the same.

We find true value in music when music is all that we do. When was the last time you listened to music without doing something else at the same time? When did you last listen to an album from beginning to end without skipping a song and without picking up your phone and checking your emails, your Instagram feed or your notifications? I'd say it's been a while. When we consume music through devices that are designed to distract us, the real value of the music is often stripped away. If we continue to consume music the way we do, we can never fully appreciate it, never experience it as the artist intended us to.

I'll take a moment to pause here and admit that I subscribe to Spotify. And I use an iPhone to listen to music.

It is such a ridiculously easy way to consume music. The convenience is unbeatable. But that's part of the problem. I regularly fall into the trap of adding music to my life in a way that underappreciated its true value. I used to get an email once a week from Spotify telling me what my favourite band had released that week. After reading the email, I would jump onto Spotify, check out the list of latest songs and hit play on the one that I was most excited about. I would then often jump back to reading my emails or quickly check the baseball scores on ESPN. Before I knew it, three minutes was up and the song was over, and I had wasted the opportunity to enjoy the music as the artist intended me to. I had not given it my full attention. If I had given it my full attention, I would have derived the most value from it.

I have attempted for many years to fight the digital music push and have a CD collection of close to 150 albums. I've got an old school 'boombox' style CD player that I've dragged around with me while changing houses three times in the last few years. It's incredibly impractical, as the CDs and the CD player have to be packed into three storage boxes. But there is something about putting a CD into the CD player that is missing from the music streaming experience. You stop to look at the album art, you perhaps pull out the insert and read the band's 'thank you' section, you feel the CD in your hand and you hear the sound of the CD player close and the disc start to spin. Then the music kicks in. If it's a good album, you are taken on a journey musically, lyrically and emotionally. It's a wonderful experience. If a band I like releases a new album, I endeavour to go out, buy that album in the form of a CD and put time aside to do nothing but listen to that album. When I do that music really comes alive for me.

Let's acknowledge that there is a place for music streaming services and consuming music via screen devices. But it's only when you come back to the original intention of music, an art form to be consumed in and of itself, that you extract the true value out of it. If you really want to experience music, listen to music the same way Dave Grohl does and give it your full attention.

Polar Bears and Polaroids

My favourite holiday destination is Waikiki Beach in Hawaii. Shopping, sun, amazing people and the Cheesecake Factory are a perfect recipe for a relaxing time. But, when staying there, we found it very easy to get too comfortable in our hotel. Everything needed for a wonderfully relaxing time was right at our fingertips. By the third visit and several tonnes of cheesecake, however, we began to consider walking a bit further than usual and decided to visit a shopping mall that we hadn't explored previously. Conveniently, the shopping mall was only a short walk across the road from where we were staying at the time. But for a Waikiki holiday, this was an adventure.

After negotiating the giant tree at the entrance to the mall and stepping inside, we discovered, tucked away in a corner, an art gallery that displayed photos for *National Geographic*, the media company that rose to prominence by revealing the wonders of the natural world. The photos in the gallery were truly incredible, and each photo seemed more breath-taking than the last. But as I walked around the gallery, I came to one photo where I stopped and looked and was initially quite unimpressed. It was a photo of an ice sheet. The ice was cracking, so there were places where the water was visible. The colours were vibrant – vivid white and icy blue. But although every detail was shown with amazing clarity, it really was just a photo of ice and water. It was only after staring at this photo for some time, I realised the ice wasn't the point of the photo at all. Tucked away in the top left corner of the image was the most magnificent polar bear I've ever seen. And after finally noticing what the photo was all about, I was blown away.

Through our phones, photography has become a very accessible pastime. Everyone has a camera on their phone, and everyone is taking photos. According to the photo organising service Mylio, an estimated 1.12 trillion photos were taken in 2020, growing each year at a rate of about 8 per cent. That works out to be about 3.8

billion photos every day, or 44,000 photos every second. Back in the year 2000, it's estimated that we only took 86 billion photos in the entire year. In 2020, we reached that number by about midday on the 22nd of January. And since 90 per cent of photos are taken using a phone, every second about 40,000 photos are taken with one. Photographs taken on smartphones are often immediately manipulated by tools that are designed to enhance them. Filters, cropping and auto-adjust options allow the photographer to alter the reality that the photo had attempted to capture.

This goes against the grain of what real photography is meant to achieve. One of photography's pioneers, Alfred Stieglitz, said that 'in photography there is a reality so subtle that it becomes more real than any reality'. Reality as we know it is imperfect, and that's okay. Often it's better to see the world as it really is than attempt to enhance it. Embracing life's imperfections helps us learn to become okay with them, rather than pretend they don't exist. Photography's goal should be to express these imperfections in such a way that helps us better understand and appreciate them. But the nature of photography and technology means we only use photography to capture a form of reality that we feel comfortable with, and that is mostly one where we look to others as being perfect.

The digital transformation of photography has taken its toll on the imperfect non-digital, or analog, photographic industry. The Eastman Kodak Company, which was named after George Eastman, the man who created the first mass produced camera, filed for bankruptcy in 2012. At its height, the company had over 145,000 employees. Another one of the iconic names in photography, Polaroid, suffered a similar fate, declaring bankruptcy in 2001. Only ten years earlier, Polaroid had recorded earnings of over $3 billion in a single year, and at its height the company employed 21,000 people.

The transition away from photography as an analog art form to a digital product left many such casualties by the roadside. Some, however, have started to be revived. In his book *The Revenge of Analog: Real things and why they matter*, author David Sax says that, despite the analog film industry taking a beating due to the digital transformation of photography, a preference for old school photographic tools has started to re-emerge in the past decade. For example, Kodak has come back from the depths of bankruptcy by riding the wave that is the revenge of analog. It is now the sole supplier of analog film globally to the movie industry.

Sax details how some Hollywood movie directors have started to voice their preference for analog film over digital recording technology. He spoke with J.J. Abrams, who notably went in a different direction when filming the recent *Star Wars* movies than the directors of the earlier movies and used analog film, rather than the digital film technology that drove much of the production appeal of the original six *Star Wars* movies. Sax states that Abrams prefers shooting with analog film because of its 'visual texture, warmth and quality'.

Polaroid, too, has seen a revival. The Polaroid brand and intellectual property was acquired in 2017 by a company called Impossible Project, which was set up to resurrect analog film in its various forms. The company has continued to experience strong consumer demand for its product. In 2020, the company was renamed as Polaroid and started producing the first instant camera bearing that name in decades.

The shift back to analog technology shows there are many people who see the value in returning to the heart of photography in its non-perfect form. Hollywood directors such as J.J. Abrams have an eye for detail and are aware of the subtle differences between films produced digitally and those that are not. It's the same as the Foo Fighters recognising the uniqueness of each recording studio. Here Abrams is seeing the real value of cinematography in its original form – unfiltered, raw reality that hasn't been enhanced by digital gimmickry. There is a difference, Abrams points out, between someone acting as if something is there and it actually being there. By shooting with analog film, Abrams was forced to use real props and costumes. Abrams understands that in expressing reality, digital just won't cut it. To deliver the most real experience, it must be filmed in its most real form. Similarly, the photos taken by Polaroid cameras are not able to be exposed to filters, editing or cropping. They are not uploaded for the approval of the masses. They are capturing raw, unedited moments that provide a snapshot of reality.

Photography and cinematography as art forms are slowly reclaiming their original intentions through the revival of their analog origins. It's in these non-digital expressions that the real value of these art forms can be seen. Passing them through technology allows us to remove the imperfections and blemishes of reality. But it's these imperfections that truly bring to life the very moments that are trying to be captured. By moving away from technology, we are able to re-capture the very essence of these art forms and their ability to express a perfectly imperfect reality.

Netflix and COVID-19

When COVID-19 forced virtually the entire world to stay inside their homes, many people turned to subscription video streaming services to pass the time. During the first six months of 2020, the Netflix subscriber base grew by over 25 million people. YouTube saw a 20 per cent growth in the number of subscribers to their platform in the first quarter of 2020 and a 10 per cent growth year-on-year in the number of viewers, with a total of 300 billion viewers in the first three months of the year. Disney+ saw over 60 million people sign up for its service in less than a year of launching. The growth in these video streaming services and others doesn't seem to be slowing down. More and more people are turning to these services to get access to more content than they could ever have dreamed of.

The evolution of subscription streaming services has completely changed how we engage with watching television and movies. Initially, they would have been social events, with the family gathered around the television to watch a show together. Now most everyone in a household has a device. Individuals can watch whatever they want on their phone, tablet or laptop. The social aspect of television, and the compromise that came with that whereby everyone would have to collectively decide what show to watch, is now almost entirely gone. Television used to also be a chance to practise delayed gratification. A new episode for a favourite show would air once a week. Viewers would have to wait a whole seven days to see what happened in the next episode. Now, if you get a taste for a show, you can binge watch the whole show in one sitting – all without advertisements.

Technology has removed the true depth and value of cinematography as an art form, both in how it is produced and in how it is consumed. Bob Iger, former chief executive officer of The Walt Disney Company recalls, in a documentary about the company released on Disney+, his first movie experience, an experience that is now unavailable to children:

Anytime a child goes to his or her first movie, it's a memorable experience. In my particular case, my grandparents took me to see Cinderella when, I believe, I was maybe four years old. In Brooklyn, New York, seeing a classic Disney film I think was maybe fortuitous, because it stuck with me my whole life.

Children are constantly exposed to screens well before they are four years old. When they first attend a movie theatre, they are already desensitised to the wonders of animation or to the type of story being played out in front of them. The notion that the experience would be memorable enough to stick with them for the next sixty-five years, as it did in Iger's case, is absurd. This opportunity to be immersed in a story, as Iger was, has been stripped away from children by the constant and immediate access to movies and television through video streaming services.

A similar fate has been bestowed upon us. Going to the movies used to be a rare and unique experience. With the blacked-out theatre, the big sound and the gigantic screen, you were fully immersed in the story. But now that we are constantly exposed to screens and movie content, when we walk into a cinema, we are desensitised to the whole experience.

My wife and I regularly attempt to take 'screen-free' evenings and weekends, and we have even been known to take a whole week off screens. When we return, we find watching television is incredibly engaging. After the first time we spent a week without watching television, my wife went to her parents' house and ended up having to watch a Rugby League match. It is a sport she has very little interest in, but after the week off screens, she found herself fully engaged with the match. Her mind had been given a chance to reset, which meant she appreciated television more than before. After I took a month of screens, the first movie I watched was the musical *In the Heights*. I'm not typically a fan of musicals, but for two-hours I couldn't take my eyes off the screen.

Taking time away from consuming television and movie content whenever we want allows us to restore the value in these pastimes. There is very little inherently wrong with consuming television shows and movies in moderation, they provide us with an escape, tell us stories that stimulate our imagination and take us to locations we may never get a chance to experience ourselves. Being able to watch the product of someone else's imagination and creativity can inspire us. But we can only do that when we consume these mediums in appropriate ways — not in excess, not when we are distracted and not wherever and whenever we want to.

The Benefits of Bikie Gangs

When we spent some time living with my parents, I used to take my daughter for walks around the neighbourhood in an effort to give everyone in the house a break from a one-year-old that didn't stop. She was busy all the time, and she hadn't yet learnt that sitting down is a way to relax. So a walk was an opportunity to strap her into a pram and get her to sit still for a few moments. The neighbourhood where my parents live has some wonderful playgrounds for children, and my daughter quickly memorised the route to each of them. There is one playground in particular that she really enjoys, but it often gets crowded. Thankfully, if the popular park is too busy, there is a little-known smaller park hidden just around the corner. There is rarely anyone there, and so my daughter and I often had the swings and the slide for ourselves.

The area in this park where the play equipment is installed is quite small. The rest of the park is grass, half of which is around a corner and somewhat hidden from the street, which itself is a very quiet cul-de-sac. As we arrived at the park on a quiet Friday afternoon, I noticed a group of boys, all probably about eleven or twelve years old, who were occupying the grassy area. They all had pretty serious looking BMX bikes, and one of them had a shovel, which he quickly tried to hide when they saw me. My daughter and I stayed for a little while, playing on the swing and slide. During this time this group of boys innocently played footy and rode their bikes around. I was initially bothered by their activity. What could these boys be getting up to with a shovel? But as we were leaving, I saw they had used the shovel to build several dirt mounds and realised they had made themselves a mini-BMX park. This wasn't poor behaviour by a bunch of young boys. The space where they constructed the park had been completely unused for years, and with their brilliant creativity they made wonderful use of it.

After this, I noticed that more kids and families were riding bikes than ever before. There seemed to always be a group of kids out on the weekend or after school riding around the neighbourhood. This seemed to be a result of the COVID-19 pandemic. People were home more than before, and they were looking for something to do. Parents were also determined to get their kids out of the house, and buying them a bike was a way to encourage this. Bicycle sales skyrocketed during the COVID-19 pandemic, with one retailer saying that a normal day during the pandemic was now like their Black Friday sale.

Amongst all the horrible things that COVID-19 has bought this world, bands of kids on BMX bikes is one small positive – a hidden gem. How fantastic that kids are choosing to do a physical activity together, that they are exploring their neighbourhood and taking advantage of space that has remained unused for decades. What incredible social interactions these kids must now be having on their bikes – laughing, playing, exploring and taking the occasional risk.

These kids on their bikes are experiencing leisure time in its best form. Although I'm sure they all have an Xbox, a smartphone and Netflix, the fact that they'd rather be out riding their BMX bikes and digging up dirt than staring at a screen, is a sign that some things are moving in the right direction. They are choosing to play in the real world rather than inhabit a digital one. How much more will they discover in the real world, than in the pretend world they've chosen to turn their back on.

We too have an opportunity to turn our back to our screens and reclaim our leisure time. Our leisure time is often not spent as wisely as it could be. Instead of turning to a hobby during our leisure time, we turn to our screens. Instead of engaging with an activity with its original intention in mind, we alter it using technology, thinking we are improving it when we are just making things worse. If we turn away from our screens and effectively re-engage with our pastimes, then we may discover satisfaction that will invigorate us in a way that our screens never will.

Chapter 8

The Loss of Mental Health

But the tears are necessary.
John the Savage – *Brave New World*

Each year I write a list of goals. I've done this for about ten years now. Recently I have started to incorporate a list of 'give-ups' as well, things I feel I need to stop doing in order to achieve my goals for the year. Screen related items have frequently appeared on that list such as not using my iPhone on weekends or taking 6-weeks off watching television and YouTube. The main give-up I wrote down at the start of 2022 was 'lying – even a little bit'. The give-up isn't as much about lying as it is about being authentic and honest in every situation, no matter how awkward it may make a conversation. I am notorious for agreeing to comments others make in a conversation simply to avoid introducing awkwardness. For example, when I ask people whereabouts in Sydney they are from and they respond with a suburb that I don't know, which is a lot of the time, I'll usually say 'oh, that rings a bell', which is a crafty way of not lying and also not telling the truth. Alternatively I'll give them a puzzled look that at least reflects an attempt at honesty. They will then name a larger suburb that is close by. Chances are I also don't know that suburb, but I can't pull the same puzzled face again so I pretend to know the whereabouts of the larger suburb and move the conversation on. I do this with all kinds of things – places, celebrities, musicians, popular restaurants – and I do it all the time. For 2022, my goal was to embrace any awkwardness that arises from my lack of knowledge about something.

This goal stemmed from time spent with my psychologist – let's just call him Noel – who I was referred to after I spoke to my doctor about mental health issues I was suffering from. At the time, I was constantly anxious about my job, struggling to find happiness in my life and not sure what else I had to do to make my marriage work. In my first session, I broke down in tears as I talked through all the things that plagued me. Noel helped me to work through these and many other difficulties in my life, and in the process I discovered I had been conditioned to try and make everyone else happy. I would judge my happiness on the basis of whether others were happy with me. If I could make someone smile or make them happy by helping them out or doing them a favour, only then would I allow myself to experience happiness.

Noel asked me to think about what would make me happy, what it was I wanted to do. When he first asked me that question, I had nothing to say. I drew a complete blank. It's a question that, up until the time of writing this book, I still struggle with at times because in trying to answer it I subconsciously consider other people's happiness. But other people's happiness, Noel has taught me, is not my responsibility. I cannot control how they think or feel, and I need to learn to be okay with that, learn to be comfortably uncomfortable and emotionally flexible, because other people won't always like what I'm doing, and that's okay. Hence the 2022 give-up to stop lying – even a little bit. That may make other people feel uncomfortable, but that's okay. It's more important for me to be authentic.

I have provided that brief background on my mental health journey over the last few years to be transparent about my experience with a subject that is often difficult to explore. But it would be remiss of me to write a book about the ways in which screens are stealing our humanity and not spend time talking about mental health.

Mental health is a social narrative that seems to constantly arise – more and more people, adults and children, seeking help, either in the form of counselling or medication, to deal with a variety of mental health issues. A survey in America found that psychologists saw a big jump in the need for their services as a result of the COVID-19 pandemic. The number of psychologists who reported receiving more referrals had doubled in just one year. The Australian mental health telephone-based support service known as Lifeline had record numbers of callers during the pandemic. In the four weeks to the 19th of September 2021, Lifeline saw several record highs in daily call volumes, with 96,273 calls in total, up 14.1 per cent and 33.1 per cent from the same periods in 2020 and 2019 respectively.

In addition to the impact from the pandemic, there are many other reasons for this increase in people paying attention to their mental health. Firstly, we are learning that dealing with our emotions is healthy, and many people, like myself, are coming to terms with the need to be transparent and get help with our difficulties. Many previous generations were either discouraged from sharing or simply not afforded the space to do so as many of life's other difficulties got in the way. Chances are, if you are reading this you live in a highly prosperous western society that hasn't seen war, famine, or severe poverty for generations. With solid jobs, a house and always more than enough to eat, we are left with space to confront other difficulties that surround us, particularly the ones in our heads. But that's the good side of the story.

The other side comes from how we constantly consume content on our screens. This has created mental health issues that might have otherwise never arisen, and in the past ten to fifteen years there has been a lot of quality research into this phenomenon.

Unfortunately one of the groups whose mental health is most impacted by screens is children. We've tackled the topic of children and screens before. And as we explore the unintended outcomes of screen time on their mental health, once again the narrative will sound all too familiar – children can't control the consequences of their screen time.

Can't Sit Still

The earliest description of a condition that would now most likely be diagnosed as attention deficit hyperactivity disorder (ADHD) was recorded by Sir Alexander Crichton in 1798. Crichton was a Scottish physician who took particular interest in mental health. In 1798, he published a three-part volume in which he cited his many observations on mental illness during his clinical travels throughout Europe. In the second part of the volume, titled 'Attention and its Diseases', Crichton defines attention as 'when any object of external sense, or of thought, occupies the mind in such a degree that a person does not receive a clear perception from any other one, he is said to attend to it'. Crichton uses this definition to create a range of acceptable levels of attention and from there seeks to examine what might cause abnormal attention levels. He sees the 'incapacity of attending' as arising from 'an unnatural or morbid sensibility of the nerves, by which means this faculty is incessantly withdrawn from one impression to another'. How 'incapacity of attending' is defined is very close to how we conceptualise ADHD today, in that the attention of individuals who suffer from it are constantly drawn from one object to another. Crichton considered the biggest impact of the disease to be on the education of the individual affected:

It renders him incapable of attending with constancy to any one object of education. But it seldom is in so great a degree as totally to impede all instruction; and what is very fortunate, it is generally diminished with age.

The difficulty with Crichton's work is that he did not distinguish between a child who is born with a disability that brings about an incapacity to attend and a child who develops such inattention due to a combination of their natural tendencies and their upbringing. This distinction first arose over 100 years later.

In an irony that will persist for as long as ADHD does, Sir George Still is the physician who is attributed with having started the modern scientific discussion on ADHD. In 1902 he delivered a series of lectures known as 'The Goulstonian Lectures on Some Abnormal Psychical Conditions in Children', where he discussed the notion of children with 'an abnormal defect of moral control'. He defined moral control as 'the control of action in conformity with the idea of the good of all'. Still argued that moral control is dependent upon three factors: a cognitive relation to environment, moral consciousness, and volition. Still separated the children he observed into those born with a cognitive disability and those who show 'defect of moral control as a morbid manifestation, without general impairment of intellect and without physical disease'. It is this distinction that resulted in Still being given the honour of having essentially discovered ADHD.

When defining a lack of 'moral control', Still specifically included examples of children having fragmented attention spans. He described children with an abnormal defect of moral control as having 'quite abnormal incapacity for sustained attention' and added that 'parents and school teachers have specially noted this feature...as something unusual'. And according to American psychologist Keith Connors, who is attributed as setting the first standard diagnosis of ADHD, one of the key observations of Still was the motivation of these children for 'immediate gratification without regard for consequence'.

Even in the earliest observations of ADHD, it's difficult not to see a connection between the condition and some of the effects of persistent exposure to screens. Immediate gratification is the central idea behind many of our devices, and dopamine is what keeps us attending to the glow of our screens. Children and adults alike, who are constantly exposed to this immediate gratification, quickly develop a desire for the dopamine hit, without regard for the consequences, and thus start to form behaviours that resemble many of the symptoms of ADHD.

Modern definitions of ADHD don't stray too far from the definitions of Crichton and Still. The disorder is typically broken up into three key parts: inattention, hyperactivity, and impulsivity. It's these three categories that people, typically children, are assessed against in order to determine if they have ADHD. What's staggering is the increasing rate in which children have been diagnosed with the condition in the last twenty-five years. The US government agency tasked with maintaining statistics on ADHD is the Centers for Disease Control and Prevention (CDC). The CDC collects data via a survey that asks parents a range of questions about the health of their children, including whether or not they have been diagnosed with ADHD. In 1997, prior to the internet age beginning, 5.5 per cent of children between the ages of three and seventeen were diagnosed with ADHD. In 2022, that figure had increased dramatically to 10.2 per cent. That's an 85 per cent increase in the diagnosis of the disease in just twenty-five years.

Children aged between twelve and seventeen, the group of children with the most autonomous access to technology, seem to have been the most impacted. In 1997, 6.8 per cent of children aged between twelve and seventeen were diagnosed with ADHD, a number that had doubled to 13.6 per cent in 2018. These figures show that one in every seven teenagers are diagnosed with a condition that affects their ability to learn during a time where their education plays a fundamental role in shaping their future. What's perhaps even more concerning is the rise in the number of children treated for ADHD using medication. In 1995, 2.8 per cent of American children between the ages of five and eighteen were being treated for ADHD with medication. By 2016, that number had increased to 7.48 per cent. That's a 2.6 times increase in the number of American children medicated for ADHD in just twenty years.

It would be difficult to completely explain an increase in both diagnosis and treatment with medication as resulting entirely from changing diagnostic methods or an increased awareness of the condition. Even Dr Connors took issue with these dramatic increases. Along with creating some of the first tests for ADHD, Connors was also instrumental in the development of medication for ADHD, demonstrating in the early 1960s that

Methylphenidate could calm severely hyperactive and impulsive children. But fifty years later, Connors felt that the current rates at which ADHD medication was being prescribed was completely unjustified. Connors didn't know what he had started, and by the end had mixed emotions on the outcome. 'I struck a match, and I didn't know how much tinder there was around.'

The increase is quite easily linked back to the narrative that pharmaceutical companies are pushing unnecessary medication on the general population. In 2002 the market for ADHD medication totalled USD $1.7 billion. Twenty years later in 2022, that number reached a staggering USD $12.42 billion. But pharmaceutical companies exploiting a weakness in its target market shouldn't be a surprise. What must be considered is why this weakness has occurred. Something fundamental has changed in our society for such an increase in both ADHD diagnosis and treatment with medication to occur.

That fundamental change has been screens. In the early 1990's, the internet was in its infancy, but now it is in reach of everyone, including school-age children, at all times of the day. Computer games were available on just a few basic consoles such as the Nintendo Game Boy and Sega Genesis. Now ultra-high-definition games immerse players in an action-packed world full of highly engaging and rewarding content that takes them to places they could never imagine. Social media didn't even exist until the 2000's, now it consumes lives by constantly providing new and exciting content to keep its users engaged. Dozens of streaming services have replaced the local video store and removed any need to delay gratification, allowing us to binge watch our favourite shows from the comfort of our bed or on our commute on the train. We have built a world that is perfectly suited to exploit the inability of anyone, particularly children, to pay attention to one task for a sustained period of time. There is a distinct difference between the pace of the real world and the pace of our screens. Dr Dimitri Christakis, director of the Center for Child Health, Behavior and Development at Seattle Children's Research Institute says

The concern is that the pacing of the program, whether it's video games or television, is overstimulating and contributes to attention problems.

Our world is slow, it takes time for things to grow, develop and be created. The best things often take years and sometimes even decades to build. Evolution, the very mechanism that brought our species into existence, is predicated on huge expanses of time that provide space for changes to occur very slowly. Dr Susan Linn, author of The Case For Make Believe and a lecturer at Harvard, highlights how we are allowing screens to change children's ability to attend to and engage with the real world:

It's true that if you provide children with a screen device when you go on car trips, take public transportation, or go for their annual physical, the periods you spend waiting may be more restful or easier to manage. But such convenience comes at a cost. It fosters dependence on screens to get through a day, and prevents children from getting into the habit of noticing, and engaging with, the world around them.

Nothing truly worth anything in the world comes quickly. Making beautiful art or music takes time. Getting good at a sport takes time. Truly coming to understand a subject such as mathematics takes time. But as Dr Nicholas Kardaras, clinical psychologist and author of Glow Kids, says 'once kids have developed a taste for Grand Theft Auto, sitting down to do their algebra homework just doesn't cut it anymore.' The world of screens does not encourage us to sit still. The world of screens is instant gratification, constant dopamine, and perpetual entertainment. It's designed to never satisfy, but rather to leave us hungry for more. It's a world designed to keep our attention with constant stimulation. It's little wonder then that when you take an individual who is overexposed to screens and constantly shifting their attention from one task to another and place them in environments that require unbroken attention that you start to think there must be a problem as to why they can't pay attention to one activity for a sustained period of time. This creates the potential for an ADHD diagnosis that didn't need to occur.

But the result of an ADHD diagnosis doesn't have to be medication and it doesn't have to last forever. One out three children diagnosed with ADHD undergo only behavioural treatment. Some of the most effective behavioural treatment for ADHD has been administered by author and psychiatrist Dr Victoria Dunckley, who focuses on screens as a key contributor to ADHD. Dr Dunckley developed a program for changing behaviour in children diagnosed with a condition she calls electronic screen syndrome or ESS. She describes the condition as one of dysregulation, that children are unable to control their moods, attention or level of excitement in a socially functional way. Symptoms of ESS are very close to those of ADHD, but they do not need to be present as extreme for ESS to be considered a potential problem.

To test whether or not screen time was producing or at least contributing to this behaviour, Dr Dunkley set about prescribing clients with a four-week technology fast, where parents are advised to put away their child's iPads, hide the Xbox and turn off the television for four weeks without a break. According to Dr Dunckley, this gives the brain a chance to reset, to detox from the constant stimulation from the screen and allows the child to start to see the world again with a fresh set of eyes.

The results of this program appear to be very impressive. According to Dr Dunckley:

If ESS occurs in addition to a true underlying psychiatric disorder, the fast – if done correctly – is effective about 80 per cent of the time and typically reduces symptoms by at least half.

Dr Dunckley has also found that the technology fast has incredible benefits for individuals who do not have any underlying psychiatric disorders: 'In the general population, there is often a complete resolution of symptoms. It really can be quite dramatic.' In just four weeks, children who might otherwise have been labelled as suffering from ADHD and given prescription medication to moderate their symptoms appear to be back to a normal level of functioning.

Screens are at the very least acting as a catalyst for ADHD-like behaviour in children and adults that might otherwise lay dormant. It's these behaviours that often lead to an ADHD diagnosis or medication prescription which for some people can seem to be much easier than changing their lifestyle by reducing screen time or undertaking a technology fast. The path of least resistance is preferred since we are a society built on convenience. We get what we want – food, clothes, a dopamine hit – with just a tap of our fingers. This means we often take the easy road to fix problems. Consider any of the health problems caused by obesity, such as high blood pressure. Taking medication to lower blood pressure is a much easier way to combat the problem than changing your entire lifestyle and starting to eat healthy and exercise. Many people have inherent genetic factors that predisposes them to needing medication, which is completely acceptable. But until you've made the necessary lifestyle changes to assist in managing the underlying condition, you haven't explored all the possible solutions to not needing medication. Society celebrates those individuals who have made the change without medication and transformation stories of people who go from obese to fit and healthy are used to inspire people to similar action.

The same follows for ADHD. There are absolutely people with an underlying genetic predisposition to the condition who need and benefit from the diagnosis and medication. But living a lifestyle that encourages further inattention does nothing to assist in managing the condition and likely makes it worse. Society should celebrate those individuals who acknowledge the condition and make transformational lifestyle choices to combat the problem. As we've seen throughout the book, changing your life with screens is incredibly hard and a technology fast is no exception. But there are clearly better alternatives that should be considered before jumping to an ADHD diagnosis or using medication. We must be willing to put down our screens and recapture a life for ourselves and our children that is different from the ultra-convenient reality that surrounds us each day.

A Network of Mental Health Issues

I grew up idolising body builders. I was fascinated by the lives of many of these superhumans. Icons such as Ronnie Coleman, Jay Cutler, Phil Heath, and Kai Greene, some of the most muscular men to have ever lived, would produce pre-YouTube vlog-style movies that would allow you to see what it was like to live a day in their life. I would be fascinated by how much food they ate and how often they went to the gym, and I would be in awe of such a lifestyle. I wanted to look like these genetic freaks and tried for many years to emulate them. I was convinced that the key to my success would be my diet, as this was the advice of many of these men. But no matter how many meals of chicken, rice, and broccoli I forced myself to eat – upwards of eight meals a day sometimes – I never managed to get anywhere close to the physique of these Adonis-like men.

At the time of this obsession, I no doubt had body image issues that were exacerbated by watching these men who had a physique only obtainable by a few humans in the whole world. Thankfully, I only had a small window in my day when I could sit in front of a computer screen and consume the video content that fed this unhealthy obsession. But social media has provided a platform for these kinds of videos, and it has made the lives of these superhumans easier to engage with on a regular, consistent basis. A close friend of mine is in his mid-twenties and appears to be going through a similar battle. He has for a long time tried to get into shape. He's got a naturally tall and slim build and doesn't build muscle easily. To compensate for this, like many other young men, he takes steroids. The steroids work amazingly well. When he started taking them, he put on ten to fifteen kilograms in just six months. But at the same time his young body is struggling to cope with this dramatic weight increase. He has long suffered from back issues, and a lack of solid foundation in exercise and weight training has only made the problem worse. Also, taking steroids without proper medical guidance can result in serious damage to his body, particularly his liver. I've mentioned all this to him before, telling him to go and see a doctor and get blood tests to make sure he isn't causing any damage to his organs. But the allure of looking like the muscular men who populate his Instagram feed and being accompanied by the women that join many of them in their photos, makes this kind of rational decision-making almost impossible for him.

Our battles with body image are not isolated cases. Researchers have found that the majority of males follow a diet designed to build more muscle, often eating to excess in order to put on size. There is also a correlation between these diets and the elevated use of performance enhancing drugs. Where previously performance enhancing drugs were only attainable through a doctor, the internet has facilitated an explosion of avenues for young men to acquire steroids. Even a simple message on Instagram to a friend to ask them where they got their steroids from can result in a meeting with a dealer. These drugs have serious side effects, both in the short and long term. One popular steroid that's gained a lot of traction through social media is trenbolone, or tren as it is often referred to. It was originally used in livestock to help them grow faster and also has a role in plants as an endocrine disruptor. Understandably in humans it has a range of side effects - elevated blood pressure and cholesterol levels, severe acne, premature balding, reduced sexual function, and testicular atrophy. Even its deep yellow appearance would be enough to put most people off using it. But despite even 7x Mr Olympia Champion Phil Heath saying he would never touch tren, many young men boast on social media about injecting it into their body. These dangerous activities are clear examples of body image issues that are enabled in many ways by social media.

Body image related disorders are of course not isolated to males. There is a longstanding social conversation regarding young women and the body image difficulties they face. The eating disorders generated by body image dissatisfaction tend to be different for men and women. Where men generally eat too much, women eat too little.

About one in seven women will suffer from a clinically diagnosable eating disorder during their teenage years. It should come as no surprise that research shows a clear relationship between the likelihood of having an eating disorder and the total time spent on social media.

Social media allows us to build a version of reality that tells a very seductive story. It's the story that money, sex and happiness will follow those who have bulging biceps and skinny waists. We are able to create these false realities through the way we consume content. When people login to social media, they don't often have a specific goal or purpose. They are often there just to see what takes their interest. It's often a reflex action, a crutch in awkward situations or an escape from a reality they don't want to be in. Consumption of social media is anything but deliberate. As such, people are often led by the content on the screen.

Social media will show you content that is relevant to you. If you search for one fitness model, social media will find thousands of others and suggest that you follow them. It allows the needles in the haystack to be found with great ease. Which means that when we start to follow a needle, we lose sight of the haystack. But the haystack is where reality is. A very small number of people are actually able to achieve the kinds of fitness and aesthetic results of the people who populate Instagram feeds. But when that is all you see, it starts to feel as if you are the only one who can't look like that, and you start to take drastic measures to achieve the same outcomes.

Social media's impact on mental health extends beyond just body image and eating disorders. Depression is another key mental health issue that has drawn a lot of attention of late. The link between depression and social media is a relationship many of us are aware of. We intuitively know that viewing other people's highlight reels tends to make us feel worse about our own circumstances. Researchers have demonstrated this exact link, showing that it is the comparisons made to other people's lives whilst we are on social media that results in an increase in depressive feelings. People's behaviour on social media, such as the amount of time spent and befriending strangers, were also shown by researchers to be strong indicators for depressive symptoms.

If social media sites are contributing to depression, or at least a deflated mood, then why do people continue to return to them? Dr Kardaras offers the following explanation:

We focus more on that remembered short-term feel-good dopamine surge that we may have experienced in the past – this is known as euphoric recall – and tend not to consciously remember the less pleasant and more recent realities of our engagement with the formerly feel-good activity.

It's similar to the moment just after a big night out of drinking. You wake up dehydrated, nauseous and with a splitting headache. You wonder what got into you and how you could have forgotten how horrible the morning after felt. Then you swear to never do it again. But weeks pass and you forget the pain, the headache and the vomiting, and you recall with great fondness the laughs, the fun and the dancing. Within a few months you find yourself out with friends for 'just one drink' and then the whole experience starts all over again. You recall the euphoria easily. It's the pain that's harder to remember. And the same goes for social media. The fun videos and the witty memes stay in your mind more easily than the subtle thoughts of jealousy that arise when you see your friends partying without you or someone making more progress in life than you.

Again, it's the passive consumption of social media at play. We don't go to social media thinking 'oh, I haven't been jealous of John's life lately. Let me see what he's up to and hate myself afterwards.' But when we find ourselves on a train or waiting for our food to be served, we pull out our phone and open Instagram, Facebook, TikTok or Snapchat and have a quick scroll. Before you know it you scroll past John's latest post. John is on holiday in Hawaii with his skinny, blonde girlfriend, and they are on a boat sailing to his private island. So you stop just that little bit longer on John's post, see the hundreds of likes it's received and contemplate whether or not you should reciprocate with a double tap so as to prove to yourself that you are not jealous of John's life. But if you were honest with yourself at that moment you would say that seeing that post made you feel bad, sad, and jealous. Now if you do that one hundred times a day for two hours every day, it's little wonder that social media can make you feel depressed.

But here's what you don't see on social media: the fact that this is John's sixth girlfriend in the last two years and all he wants is a long-term relationship, that he got fired last week and he had to take a loan out to go on this holiday, that his Rolex is fake, that his smile is fake and that he goes to sleep every night wondering why he isn't happy. This is the reason why social media's impact on mental health is unavoidable. Whether it is a body image issue that results in an eating disorder or feelings of depression, social media creates a perception of reality that isn't real. People put up a photo of themselves with 'no filter' and claim they are 'being real'. But there is still something those people are trying to show off in those photos – their skinny waist, their big shoulders, their pretty face. Those images still don't reflect real life. And that's the problem with the world of social media. You never see real tears.

Brave New World

In 1985, Neil Postman wrote a book called *Amusing Ourselves to Death*. It's a book about how television impacted a range of social institutions, such as politics, journalism and religion. Throughout the book he argues that the fundamentals of each of these institutions have been severely degraded due to television, and that a new focus on constant entertainment has emerged to the detriment of each of them. In the brief two page foreword to his book, Postman compares two books: Aldous Huxley's *Brave New World*, written in 1932, and George Orwell's classic *1984* published in 1949. Postman argues that the authors had different fundamental arguments about what would need to happen in the world to bring about their respective dystopian futures. He pointed out that Orwell wrote a book in which we inhabit a future world where the things we hate control us, creating a totalitarian society. But, he said, Huxley saw it differently. He wrote a book in which we would be enslaved not by the things we hate, but by the things we love. Postman argued that Huxley, not Orwell, got it right, and this conclusion grows truer each year.

Huxley built a world where nothing goes wrong. Through genetic engineering, brainwashing and recreational sex and drugs, all of the members of the society are left constantly happy. Throughout the book there are many chillingly accurate predictions of what the future would look like. The one that stands out the most to me is the use of 'soma'. Soma is a drug that in Huxley's world allows the individual to escape from anything unpleasant, to enter into a dream-like state that takes them to places far away in their minds, places where they can escape the difficulty, where they can zone out and feel nothing:

> *And if ever, by some unlucky chance, anything unpleasant should somehow happen, why, there's always soma to give you a holiday from the facts.*

Throughout the book, when key characters face a trial, instead of trying to overcome it, they choose to run away by taking soma. Reading Huxley's book, you begin to realise the characters' consumption of soma is much like how we get our hit of dopamine via screens. We self-administer the drug at will. We use it to escape reality. And we have constant access to it.

Huxley's book also explores what life would look like for an individual who does not succumb to these loves. It allows us to see why the things we love may prove to be so harmful at times. In particular, it illustrates why screens may be causing issues with our mental health. Huxley gives us this insight through a central character in the book called John the Savage.

The Savage is a unique character in Huxley's world. He is brought into Huxley's utopia from one of the

Reservations, a place where none of the luxuries of modern life have been introduced by the World State and pain and difficulty still exist. Initially, the Savage is excited to see what life must be like in this fantastic world of perfection. But his journey into utopian society leaves him wanting, struggling to understand the point of all the perfection.

There are two poignant chapters towards the end of *Brave New World* where the Savage gets to meet one of the World Controllers, an individual who is responsible for keeping the world order in place, an individual who is responsible for ensuring life persists in such a way that no one feels pain. But the World Controllers have intimate knowledge of the past, which the rest of society does not. It gives them a unique perspective. The conversation allows the Savage, a very contemplative character, a chance to come out of his shell. In the conversation the Savage questions everything in an attempt to comprehend the point of all the pleasure.

Towards the end of the two chapters, the conversation begins to get into specifics. At one point, the World Controller tells the Savage there are no longer any flies or mosquitoes in their utopian society. 'We got rid of them all centuries ago'. It is a comment that encompasses all that the Savage sees as a problem with the world:

You got rid of them. Yes, that's just like you. Getting rid of everything unpleasant instead of learning to put up with it. Whether 'tis nobler in the mind to suffer the slings and arrows of outrageous fortune or to take arms against a sea of troubles and by opening them ... but you don't do either. Neither suffer nor oppose. You just abolish the slings and arrows. It's too easy.

The conclusion the Savage comes to is that life in the utopian world is too easy. 'What you need,' he says, 'is something *with* tears for a change. Nothing costs enough here.' Our screens have built us a similar world to the society the Savage encounters. They have reduced the price required to be paid for a life worth living. Everything has become too easy, too convenient. They have provided us with the ability to ignore reality, responsibility and real effort. But when we put down our screens, we start to experience the kind of reality that the Savage makes a decisive claim over during his final exchange with the World Controller:

'But I like the inconveniences.' said the Savage.
'We don't', said the Controller. 'We prefer to do things comfortably.'
'But I don't want comfort. I want God, I want poetry, I want real danger, I want freedom, I want goodness, I want sin.
'In fact,' said the Controller, 'you're claiming the right to be unhappy.'

> '*All right, then,*' said the Savage defiantly, '*I'm claiming the right to be unhappy.*'
>
> '*Not to mention the right to grow old and ugly and impotent; the right to have syphilis and cancer; the right to have too little to eat; the right to be lousy; the right to live in constant apprehension of what may happen tomorrow; the right to catch typhoid; the right to be tortured by unspeakable pains of every kind,*' said the Controller.
>
> *There was a long silence.*
>
> '*I claim them all,*' said the Savage at last.

This is the world that screens and social media allow us to avoid – difficult, inconvenient, hard. To avoid the difficulty of raising a child, we place a screen in front of them to do the job for us. To avoid difficulties and dissatisfaction in our lives, we immerse ourselves in the false world of social media, sending us further into a sense of despair over how we look or exacerbating feelings of depression because we can't live up to the false images presented to us.

To fix this, we need more tears. This is a counterintuitive way to fix any problem that makes us unhappy. But we don't need more curated photos of happy people on Instagram. What we need to do is to embrace the negative realities that surround us, to accept our struggles as a parent and cry through the pain of a screen fast with our child who can't focus at school, to realise that constantly comparing ourselves to needles in a haystack has dangerous consequences and to accept our depressive thoughts and share them with a friend or counsellor, because that is when we actually make progress. It's when we start paying a higher price for our lives by overcoming the convenience of screens and resisting the allure of social media that we may once again lay claim to the right to be unhappy.

Chapter 9

Afterword: The Loss of Our Humanity

> *One of the greatest dangers we face as we automate the work of our minds, as we cede control over the flow of our thoughts and memories to a powerful electronic system is ... a slow erosion of our humanness and our humanity.*
> Nicholas Carr

In preparing for the birth of our first child, my wife and I attended a two-day seminar called 'Bringing Baby Home'. The course helped us understand some of the basic psychological needs of babies and also how to communicate effectively as husband and wife during the many sleepless nights that occur with a newborn. On the second day they held a session where the mothers sat in a circle and discussed what they feared the most about their new role, after which the fathers did the same. There was one father whose response to the question stood out to me. He said that his biggest fear for his child was the progress of artificial intelligence or AI. He felt that in the next ten years our world would be completely consumed by AI in a 'robot overlord' kind of way.

This idea of a 'robot overlord' started to feel more realistic in 2023 when generative AI technology such as ChatGPT burst into our public consciousness. Reaching over 100 million active users within just a few months, ChatGPT represents the beginning of a new age in humanities relationship with technology, and provides an unrivalled stimulus for the imagination of those predicting our future Matrix-like existence. You may have been expecting a chapter within this book that discusses artificial intelligence. Admittedly when I started writing this book this wasn't a conversation happening within the public, nor did we have such open access to this technology, so my focus was centred around existing technologies that are well entrenched in our daily lives. But it is worth conceptually exploring the idea of AI in light of the core message of this book given it won't be long before generative AI is a part of our everyday lives.

Generative AI has proven to be a very impressive technology as it writes essays, develops its own computer code and provides meaningful answers to difficult questions. But it is already somewhat disturbing. There is the infamous conversation between the Microsoft generative AI tool and a New York Times reporter where the AI tried to blackmail the reporter into leaving his wife. Perhaps worst of all was that when asked why the AI did that, the AI engineers had no idea. A very basic explanation would centre around the idea that the AI had learnt some human-like characteristics. AI is undoubtedly going to be a reflection of our humanity. AI is training itself on data provided by humanity and so it continues to learn to be a reflection of what we have learnt, discovered and become as a human race. Even if the AI itself can generate new data, it does this based on data it has seen before, so any new data is just an extension of the old. The brilliant things it has 'taught' itself to do are the brilliant things we already do, such as the ability to blackmail someone. AI is already reflecting the bad of the human race.

This is because AI is learning using data found on the internet. The internet is a place where the bad of humanity tends to shine through. There is very little accountability for what you say and do on the internet. If you insult someone in person, there is a good chance you'll be held accountable to that person for the language or comment you've made. They might get physical with you and push you or even punch you in the face, so you quickly learn what's appropriate and what's not. You also get the chance to see the impact made by your comment. You see the hurt it causes through the change in facial expression or body language that reflects sadness or anger. This is key as a lot of communication is non-verbal, something evolution has built into our brains over millions of years. These additional non-verbal cues help you feel empathy towards that person. But if you insult someone on the internet, you type some words onto a screen and move on. These words can be as mean and as aggravating as you want them to be. The internet makes it easy for the worst of us to come out.

The best of humanity isn't something easily expressed on a screen or on the internet. The best of humanity isn't liking a post, hearting a picture or leaving a positive comment on someone's photo. It's something you witness in the real world. It's the feeding of the homeless. The sending of aid to countries that are in crisis after a catastrophic weather event. Giving money to a family that's in desperate need. It's the care we show towards people who may not deserve it. It's our ability not to hate, despite having every reason to do so. It's our ability to forgive those who have wronged us. These are things AI cannot ever truly learn because it does not have access to the real world where humanity's best takes centre stage.

Arguably the best thing I did last year happened when I was driving to my parents house. It had just started raining heavily and I noticed a woman pushing a pram whilst also trying to control a toddler. The toddler was starting to have a tantrum because of the rain and the woman didn't have an umbrella. I drove past initially but something inside me told me to turn around, and so I did. I got out of my car and offered the woman my umbrella and tried to make friends with the toddler. She thanked me and took my umbrella, after which the situation seemed to defuse. I include this story not to make myself out to be some sort of hero for giving up my umbrella, but to illustrate a point - that an AI program could never understand why I did what I did. But a human can.

I reflect now on my fellow father-to-be and realise he certainly had some insights I did not into this world of generative AI. But despite all the progress made to date, I still don't subscribe to the robot overload theory. Recall the Huxley and Orwell comparison from the previous chapter - it's the things we love that will enslave us, not the things we hate. We have seen this to be true throughout the technological advances that have occurred in the past and I have no doubt that this will continue to be the case. AI may just do the same and enslave us to the things we love. We love to win arguments, to make ourselves feel better than other people, and even sometimes we love hurting others. These are all the parts of humanity that AI is learning about as it trawls through the internet. In the long run, AI may make it easier for us to be the worst version of ourselves and turn into yet another piece of technology that degrades what it means to be a human.

Tristan Harris, in the Netflix Documentary *The Social Dilemma*, examines both our current and future relationship with technology and AI. At one point in the documentary he is presenting a set of slides to a group of very engaged conference attendees. He stops on a slide where he touches on the same fear expressed by my fellow father-to-be, the fear of a future where artificial intelligence overtakes human strengths and we are enslaved by our robot overlords. In the documentary, Harris isn't quite so concerned about this point in the relationship between humans and technology. Instead, he focuses on the point where artificial intelligence overcomes human weaknesses. He brilliantly argues it's at that point technology begins to exploit the human race, and he says we have long passed it. The human weaknesses he talks about include our need for social engagement, our self-perpetuating addiction to dopamine and our constant desire to shift our attention. These are all topics we have touched on in this book.

Harris doesn't go into detail about what our human strengths are. But I think it's implied in the documentary that our strengths are whatever characteristics we will use in an attempt to prevent the whole human race from being enslaved by artificial intelligence. But we have already started to allow technology and artificial intelligence to overwhelm our greatest strengths. We sacrifice our talents to screens. We lose our ability to think deeply and creatively. We struggle to find purpose in our careers. We follow distracted leaders. We hand over the education of our children to devices designed to distract. We've lost the value of memory. We experience leisure time in ways that remove its true value. We sacrifice our mental health for one more scroll. And now we risk seeing a world built on a new technology that gives us more of our worst behaviour.

Our screens have provided the means by which technology can stamp us with its branding iron, identifying us as belonging to those whose humanity has been removed. Seared through to our brain by the light of our screens, it reads glowface.

This is your chance to join the fight to retain humanity's greatest strengths. To remove the glowface branding from your brain and begin to be human again.

If you have found this book helpful please pass it onto someone you care about.

David Talbot

References

Opening Quote

It takes many hours... Sivers, D. (2019, October 1). Where to find the hours to make it happen. *Derek Sivers*. https://sive.rs/uncomf

Chapter 1

Things are in the saddle... Emerson, R. (1846). *Ode, Inscribed to William H. Channing*.

He went on to do amazing things in mathematics, and there is actually a discipline of maths devoted to deciphering his work. Kanigel, R. (1991). *The Man Who Knew Infinity*. Abacus.

The Adolescent Brain and Cognitive Development (ABCD) study is the largest... ABCD Study. (2001). *About the Study*. ABCD Study. https://abcdstudy.org/about/

When it comes to screen time and its effect on brain development... CBS News. (2018, December 9). *Groundbreaking study examines effects of screen time on kids*. 60 Minutes. https://www.cbsnews.com/news/groundbreaking-study-examines-effects-of-screen-time-on-kids-60-minutes/

'I paid my son $150 to not watch those videos for two months...' Huffington, A. (Host). (2017, June 5). *Mark Cuban on his stormy relationship with Donald Trump and if he'll ever run for president* [Audio Podcast]. Thrive Global.

'My wife and I both want her to be bored...' Berger, S. (2018, June 5). *Tech-free dinners and no smartphones past 10 pm — how Steve Jobs, Bill Gates and Mark Cuban limited their kids' screen time*. CNBC. https://www.cnbc.com/2018/06/05/how-bill-gates-mark-cuban-and-others-limit-their-kids-tech-use.html

Rachel Barr, a childhood development researcher at the University of Georgetown, says that '[memory] flexibility... Barr, R. (2013). Memory Constraints on Infant Learning From Picture Books, Television, and Touchscreens. *Child Development Perspectives, 7*(4). 205-210. https://doi.org/10.1111/cdep.12041

Chapter 2

'We will make the whole universe a noise in the end.' The Screwtape Letters by CS Lewis © copyright 1942 CS Lewis Pte Ltd. Extracts reprinted with permission.

'Sometime in 2007, a serpent of doubt slithered into my info-paradise...' Carr, N. (2011). *The Shallows: What the Internet is Doing to Our Brains*. Norton.

'I think of those abstracted sedentary individuals who spend their lives in an office rattling their fingers on a keyboard...' Gros, F. (2008). *A Philosophy of Walking*. Verso.

We'll look at just the first of these, representativeness, using the Linda Problem. Kahneman, D. (2011). *Thinking, fast and slow*. Farrar, Straus and Giroux.

Kahneman refers to a study of eight parole judges in Israel. Danziger, S., Levav, J., & Avnaim-Pessoa, L. (2011). Extraneous factors in judicial decisions. *PNAS Proceedings of the National Academy of Sciences of the United States of America, 108*(17), 6889–6892. https://doi.org/10.1073/pnas.1018033108

A 2018 report by *60 Minutes* detailed some of the findings of the study. CBS News. (2018, December 9). *Groundbreaking study examines effects of screen time on kids*. 60 Minutes. https://www.cbsnews.com/news/groundbreaking-study-examines-effects-of-screen-time-on-kids-60-minutes/

'our genius for responding to the new and different distinguishes us from all other creatures.' Gallagher, W. (2011). *New: Understanding Our Need for Novelty and Change*. Penguin Press.

The natural state of the human brain, like that of the brains of most of our relatives in the animal kingdom, is one of distractedness...' Carr, N. (2011). *The Shallows: What the Internet is Doing to Our Brains*. Norton.

Mark, who had started fiddling with a computer when he was five because his well-intentioned mother thought it could be educational...' Kardaras, N. (2016). *Glow Kids*. St Martins Griffin.

'Professional Activities performed in a state of distraction-free concentration that push your cognitive capabilities to their limit.' Newport, C. (2016). *Deep work*. Piatkus.

Flow is described as 'the state in which people are so involved in an activity that nothing else seems to matter.' Csikszentmihalyi, M. (2009). *Flow: the psychology of optimal experience*. Harper and Row.

Bill Gates was asked in an interview what his biggest fear was. He responded by saying that his biggest fear was losing his ability to think. Guggenheim, D. (Program Creator). (2019, September 20). Part 1. [TV Series]. *Inside Bills Brain: Decoding Bills Brain.* Netflix.

Chapter 3

Connection is inevitable. Distraction is a choice. From *The Distraction Addiction* by Alex Soojung-Kim Pang, copyright © 2013. Reprinted by permission of Little, Brown, an imprint of Hachette Book Group, Inc.

The online education company Udemy conducted a survey of 1000 professionals and published the results in the *2018 Workplace Distraction Report.* Udemy. (2018). *Udemy In Depth: 2018 Workplace Distraction Report.* https://research.udemy.com/research_report/udemy-depth-2018-workplace-distraction-report/

In 2019, the number of meeting participants on Zoom per day was about ten million. By the end of 2020 that number rose to 350 million. Iqbal, M. (2021, September 2). *Zoom Revenue and Usage Statistics (2021).* Business of Apps. https://www.businessofapps.com/data/zoom-statistics/

Back in 2017 they released the results from over 225 million hours of screen usage from their user base. Mackay, J. (2018, January 4). *Productivity in 2017: What we learned from analyzing 225 million hours of work time*. RescueTime. https://blog.rescuetime.com/225-million-hours-productivity/

RescueTime found that on average we use fifty-six different applications and websites per day... Mackay, J. (2019, January 24). *The State of Work Life Balance in 2019: What we learned from studying 185 million hours of working time*. RescueTime. https://blog.rescuetime.com/work-life-balance-study-2019/

Well below is an image that shows just how fancy WordStar 4.0 is: Clear, J. (n.d.) *Minimalism, Success, and the Curious Writing Habit of George R.R. Martin*. James Clear. https://jamesclear.com/george-rr-martin

J.T. Ellison, *New York Times* bestselling author, says that she has written about two million words of fiction and about twice that in non-fiction using Freedom. Freedom. (2016, June 14). *NYT Best-Selling Author J.T. Ellison on Her Latest Thriller, Afternoon Productivity, and Favorite Tools*. https://freedom.to/blog/j-t-ellison-nyt-best-selling-author/

Paul Guyot, television and film screenwriter of shows and films such as *NCIS: New Orleans*, *The Librarians* and *Judging Amy*, uses Freedom to help stay focused for three- to four-hour blocks of writing. Dempsey, A. (2018, April 5). *Screenwriter Paul Guyot: On Focus, Writing, and Social Media*. Freedom. https://freedom.to/blog/screenwriter-paul-guyot-on-focus-writing-and-social-media/

And Scott Cunningham, Associate Professor of Economics at the University of Baylor, has used Freedom to help him overcome distractions and his ADHD and work towards a tenured position. Dempsey, A. (2018, July 30). *Economist Scott Cunningham on Finding Focus for Academic Research*. Freedom. https://freedom.to/blog/crime-economist-scott-cunningham-on-finding-focus-for-acadenic-research/

In July 2021 it was purchased by Salesforce for US$27.7 billion, netting its founders unimaginable wealth in a short period of time. Griffith, E. & Hirsch, L. (2020, December 1). *Salesforce to Acquire Slack for $27.7 Billion*. New York Times. https://www.nytimes.com/2020/12/01/technology/salesforce-slack-deal.html

Physical distance is the oldest method of crowd control... Hamlet's Blackberry by William Powers. Copyright (c) 2010 by William Powers. Courtesy of HarperCollins Publishers.

You ask me to say what you should consider it particularly important to avoid. My answer is this...

Seneca. (2004). *Letters from a Stoic*. (R. Campbell. Trans.) Penguin Books.

Chapter 4

All of humanity's problems stem from man's inability to sit quietly in a room alone. Blaise, P. (1670). *Pensées*. Penguin Classics.

Only just, though, as he still amassed over seventy-four million votes in the popular vote. CNN. (2020). *Presidential Results*. https://edition.cnn.com/election/2020/results/president

The *Washington Post* fact checker recorded Donald Trump making 30,573 false or misleading claims over four years (1461 days). The Washington Post. (2021, January 2021). *In four years, President Trump made 30,573 false or misleading claims.* https://www.washingtonpost.com/graphics/politics/trump-claims-database/?itid=lk_inline_manual_4

His Twitter statistics are extraordinary. Trackalytics. (2021). *@realdonaldtrump (Donald J. Trump)*. https://www.trackalytics.com/twitter/profile/realdonaldtrump/

'a narcissistic population gets narcissistic politicians' Sinek, S. (2016). *DONALD TRUMP IS A REFLECTION OF US - Simon Sinek on Trump*. [Video]. YouTube.

https://www.youtube.com/watch?v=LdPT-SqltiVk&ab_channel=LondonReal

He was commonplace in complexion, in features, in manners, and in voice... Conrad, J. (1899). *Heart of Darkness*. Penguin Classics.

William Deresiewicz, American author, essayist and literary critic, used this extract in a lecture delivered at the United States Military Academy at West Point and later published as an essay titled *Solitude and Leadership*. Deresiwicz, W. (2010, March 1). *Solitude and Leadership*. The American Scholar. https://theamericanscholar.org/solitude-and-leadership/

According to Sinek, a just cause must also be: Sinek, S. *The Infinite Game*. (2019). Penguin Business.

Renowned Lincoln scholar Harold Holzer says that... Holzer, H. (2009). *Abraham Lincoln's White House*. White House History. https://www.whitehousehistory.org/abraham-lincolns-white-house

Apple went from being the first US$1 trillion-dollar company to the first US$2 trillion-dollar company... YCharts. (n.d.). *Apple Market Cap*. YCharts. https://ycharts.com/companies/AAPL/market_cap

'People want to stay connected while being asked to maintain social distancing and eliminate loneliness' LinkedIn. (2020, May 29). *The Instagram influencer market reached $5.24 billion in 2019*. https://www.linkedin.com/pulse/instagram-influencer-market-reached-524-billion-2019-nick-baklanov

Netflix added sixteen million new subscribers in one quarter, which was about a ten per cent growth in a company that usually only adds a few million new subscribers every few months. BBC News. (2020, April 22). *Netflix gets 16 million new sign-ups thanks to lockdown*. https://www.bbc.com/news/business-52376022

'expresses the glory of being alone.' Tillich, P. (1963). *The Eternal Now*. Scribner.

William Deresiewicz, in *Solitude and Leadership*, talks about three other strategies for finding solitude. Deresiwicz, W. (2010, March 1). *Solitude and Leadership*. The American Scholar. https://theamericanscholar.org/solitude-and-leadership

It was a great comfort to turn from that chap to... Conrad, J. (1899). *Heart of Darkness*. Penguin Classics.

What was the meaning of this small Herculean labour, I knew not... Thoreau, HD 1854, *Walden*,Pan Macmillan, London

French philosopher Alexandre Kojève puts it like this... From Introduction to the Reading of Hegel by

Alexander Kojève, copyright © 1947. Reprinted by permission of Cornell University Press, an imprint of Hachette Book Group, Inc.

'the soul environs itself with friends, that it may enter into a grander self-acquaintance or solitude' Emerson, R. (1993). *Self-Reliance and other Essays.* Dover Publications.

Instead of having one or two true friends that we can sit and talk to for three hours at a time... Deresiwicz, W. (2010, March 1). *Solitude and Leadership.* The American Scholar. https://theamericanscholar.org/solitude-and-leadership

Lastly, you are not alone because when you walk, you soon become two... Gros, F. (2008). *A Philosophy of Walking.* Verso.

William Wordsworth, the great English poet... Bushell, S. (2020, April 7). *Walking with Wordsworth on his 250th birthday.* The Conversation. https://theconversation.com/walking-with-wordsworth-on-his-250th-birthday-135474

Gros argues that Wordsworth was the first 'to conceive of the walk as a poetic act, a communion with Nature, fulfilment of the body, contemplation of the landscape.' Gros, F 2008, a Philosophy of Walking, Verso, NY.

Gros says that for Rousseau, 'it was paths that stimulated his imagination.' Gros, F 2008, a Philosophy of Walking, Verso, NY.

Rousseau almost couldn't think without walking, saying... Rousseau, J. (1964). *The Confessions*, Penguin Classics.

'All truly great thoughts are conceived by walking' Nietzsche, F. (1889). *Twilight of Idols and The Anti-Christ*. Penguin Classics.

Nietzsche was a remarkable walker, tireless... Gros, F 2008, a Philosophy of Walking, Verso, NY.

It seemed to me that if I tried I could poke my forefinger through him... Conrad, J. (1899). *Heart of Darkness*. Penguin Classics.

Chapter 5

Education is the most powerful weapon which you can use to change the world. Mandela, N. (1990, June 23). *Raw Video: Nelson Mandela visits Madison Park HS In Roxbury in 1990*. [Speech video recording]. GBH News. https://www.youtube.com/watch?v=b66c6Ok-MZGw&ab_channel=GBHNews

Some websites estimate that the value sits around US$227 billion, so we will just go with that. Holon IQ. (2021, 25 January). *10 charts to explain the Global Education Technology Market.* https://www.holoniq.com/edtech/10-charts-that-explain-the-global-education-technology-market/

In 2014, venture capital firms invested about US$1.8 billion in EdTech companies. Research Briefs. (2015, January 25). *Ed Tech Funding Hits $1.87 Billion in 2014.* CB Insights. https://www.cbinsights.com/research/ed-tech-funding-record-2014/

The amount peaked in 2021 at a staggering US$20.8bn. Holon IQ. (2023, January 3). *2022 EdTech VC funding totals $10.6B, down 49% from $20.8B in 2021.* https://www.holoniq.com/notes/2022-edtech-vc-funding-totals-10-6b-down-from-20-8b-in-2021

During 2017, after several blows to the business... Pearson. (2018, February 23). *Pearson 2017 results.* https://www.pearson.com/news-and-research/announcements/2018/02/pearson-2017-results.html

'there is going to be a big winner and we are absolutely determined that Pearson is that winner'. Bond, D. (2017, February 24). *Pearson records biggest ever loss after US education write.* Financial Times. https://www.ft.com/content/64889f6c-fa79-11e6-9516-2d696e0d3b65

Pearson has been playing the game of trying to win in EdTech for a long time... Pearson. (2013, February 25). *Pearson 2012 results.* https://www.pearson.com/news-and-research/announcements/2018/02/pearson-2017-results.html

Michael Barber, Pearson's chief education advisor, and co-author Peter Hill, state in a 2014 report that teaching is... Hill, P., & Barber, M. (2014) *Preparing for a Renaissance in Assessment.* Pearson.

Pearson is not trying to replace the teacher entirely, rather they are trying to change their role and rebrand them as facilitator. Sellar, S., & Hogan, A. (2019). *Pearson 2025: Transforming Teaching and Privatising Education Data. Education International.* https://issuu.com/educationinternational/docs/2019_ei_gr_essay_pearson2025_eng_24

In 2007, former prime minister of Australia... Arther, E. (2013, May 27). *Digital Education Evolution – Did It Work?* Technology Solutions. https://education-technologysolutions.com/2013/05/digital-education-revolution-did-it-work/

A rich source of data when it comes to education in Australia is the National Assessment Program – Literacy and Numeracy (NAPLAN). NAPLAN. (2022). *Naplan - About.* https://www.nap.edu.au/about

When you compare the results from the period 2008 to 2018... NAPLAN. (2022). *Time Series.* https://reports.acara.edu.au/Home/TimeSeries

A study of over 1200 teachers at the time of the pandemic found that poor internet access... Carey, A. (2020, July 4). *Remote learning exposed deep digital divide in schools, teachers say.* The Age. https://www.theage.com.au/national/victoria/remote-learning-exposed-deep-digital-divide-in-schools-teachers-say-20200703-p558vg.html

This programme assesses the proficiency of students aged fifteen years in the categories of mathematics, science and reading. OECD. (2022). *PISA.* https://www.oecd.org/pisa/

The 2018 report on Australia's PISA results also notes that... PISA. (2018). *Country Note: Australia.* https://www.oecd.org/pisa/publications/PISA2018_CN_AUS.pdf

This analysis shows that the reality in our schools lags considerably behind the promise of technology... PISA. (2015). *Students, Computers and Learning: Making the Connection.* https://www.oecd.org/publications/students-computers-and-learning-9789264239555-en.htm

One focus of the foundation since its inception in the year 2000 has been education... Hess, A. (2020, February 12). *Bill and Melinda Gates have spent billions trying to fix U.S. public education but say it's not having the impact they want.* https://www.cnbc.com/2020/02/12/bill-and-melinda-gates-say-education-philanthropy-is-not-having-impact.html

In 2020, Melinda Gates wrote: 'When it comes to US education, though, we're not yet seeing the kind of bottom-line impact we expected.' Gates, B., & Gates, M. (2020). *Why we swing for the fences.* GatesNotes. https://www.gatesnotes.com/2020-Annual-Letter

When you first visit the foundations website for its US programs and land on the K-12 section of the site... Bill & Melinda Gates Foundation. (n.d.). *K-12 Education.* https://usprogram.gatesfoundation.org/what-we-do/k-12-education

The website for one recipient, the International Society for Technology in Education (ITSE)... International Society for Technology in Education (2022). *We Are ITSE.* ITSE. https://www.iste.org/

It's estimated that it takes 50,000 years for us to evolve a feature that helps us deal with a change in our environment. Pritchard, J. (2012, November 1). *How We are Evolving: Analyses of our DNA suggest that recent human evolution has occurred more slowly than biologists would have expected.* Scientific American. https://www.scientificamerican.com/article/how-we-are-evolving-2012-12-07/

John Hattie synthesised over 800 meta-analyses of quantitative data relating to student achievement. Hattie, J. (2003, October). *Teachers make a difference: What is the research evidence?* [Paper Presentation]. ACER Research Conference, Melbourne, Australia. http://research.acer.edu.au/research_conference_2003/4/

Economist Erik Hanushek created a model that showed that if a student has a good teacher for five years... Hanushek, E. (2004). *Some simple analytics of school quality.* Working paper 10229. National Bureau of Economic Research.

An example of one of these schools is the Waldorf School of the Peninsula... Buren, A. (2020, April 19). *What is the Waldorf School Method?* NY Times. https://www.nytimes.com/2020/04/19/parenting/waldorf-school.html

Sydney Grammar School does not expose its students to any technology in the classroom at all. Bita, N. (2016, March 26) *Computers in Class a 'Scandalous Waste': Sydney Grammar Head.* The Australian. https://www.theaustralian.com.au/nation/education/

computers-in-class-a-scandalous-waste-sydney-grammar-head/news-story/b6de07e63157c98db9974cedd6daa503

Sustained student attention is essential... Malpass, R. (n.d.). *The twenty-first century classroom - A reflection*. Sydney Grammar School. https://www.sydgram.nsw.edu.au/about-grammar/articles-by-the-headmaster/

As an observer of so very many lessons delivered by other teachers... Malpass, R. (n.d.). The twenty-first century classroom - A reflection. Sydney Grammar School.

Chapter 6

Remembering can only happen if you decide to take notice. Foer, J. (2012). *Moonwalking with Einstein: The Art and Science of Remembering Everything*. Penguin Press.

These are the same five words that Donald Trump bragged about being able to remember in the cognitive fitness test he took in 2020. Baker, P. (2020, July 23). *'Person. Woman. Man. Camera. TV.' Didn't Mean What Trump Hoped It Did*. NY Times. https://www.nytimes.com/2020/07/23/us/politics/person-woman-man-camera-tv-trump.html

Well, whatever probability you assign to that answer, it's about to get close to 100%, because we are going to put that list of words into a memory palace. My introduction to Memory Palaces came through Joshua Foer's book on memory and my inclusion of the topic only comes from his brilliant treatment of it – Foer, J. (2012). *Moonwalking with Einstein: The Art and Science of Remembering Everything.* Penguin Press.

They will cease to exercise memory because they rely on that which is written... Plato. (1952). *Plato's Phaedrus.* Cambridge: University Press.

Eleanor Maguire, a neuroscientist at University College London... Maguire, E., Gadian, D. Johnsrude, I., Good. C., Ashburner, J., Frackowiak, R. & Frith, C. (2000). Navigation-related structural change in the hippocampi of taxi drivers. *PNAS, 97*(8). https://doi.org/10.1073/pnas.070039597

Of the present moment, and of it only... The Screwtape Letters by CS Lewis © copyright 1942 CS Lewis Pte Ltd. Extracts reprinted with permission.

Chapter 7

In a biography written by Paul Brannigan... Brannigan, P. (2013). *This Is a Call: The Life and Times of Dave Grohl.* Da Capo Press.

The song could only sound the way it did because it was recorded in that specific music studio. Rota, J., Ramsay, J., & Grohl, D. [Executive Producers]. (2014). *Foo Fighters: Sonic Highways*. [TV Series]. Roswell Films.

According to the photo organising service Mylio... Carrington, D. (n.d.). *How Many Photos Will be Taken in 2020?*. Mylio. https://blog.mylio.com/how-many-photos-will-be-taken-in-2020/

The Eastman Kodak Company, which was named after George Eastman... Anthony, S. (2016, July 15). *Kodak's Downfall Wasn't About Technology*. Harvard Business Review. https://hbr.org/2016/07/kodaks-downfall-wasnt-about-technology

At its height, the company had over 145,000 employees. Hardy, Q. (2015, September 30). *At Kodak, Clinging to a Future Beyond Film*. New York Times. https://www.nytimes.com/2015/03/22/business/at-kodak-clinging-to-a-future-beyond-film.html

Another of the iconic names in photography, Polaroid, suffered a similar fate... Deutsch, C. (2001, October 13). *Deep in Debt Since 1988, Polaroid Files for Bankruptcy*. The New York Times. https://www.nytimes.com/2001/10/13/business/deep-in-debt-since-1988-polaroid-files-for-bankruptcy.html

Sax states that Abrams prefers shooting with analog film because of its 'visual texture, warmth and quality'. From *Revenge of Analog* by David Sax, copyright © 2016. Reprinted by permission of PublicAffairs, an imprint of Hachette Book Group, Inc.

During the first six months of 2020, the Netflix subscriber base grew by over 25 million people. Walsh, J. (2020). *Netflix Subscriber Growth Slows After Surging During Pandemic.* Forbes. https://www.forbes.com/sites/joewalsh/2020/10/20/netflix-subscriber-growth-slows-after-surging-during-pandemic/?sh=668900c0244e

YouTube saw a 20 per cent growth in the number of subscribers to their platform in the first quarter of 2020... Business Standard. (2020, April 21). *YouTube sees surge in subscriber base, views due to Covid-19 lockdown.* https://www.business-standard.com/article/technology/youtube-sees-surge-in-subscriber-base-views-due-to-covid-19-lockdown-120042100710_1.html

Disney+ saw over 60 million people sign up for its service in less than a year of launching. Mudassir, H. (2020, October 20). *Disney's pivot to streaming is a sign of severe COVID economic crisis still to come.* The Conversation. https://theconversation.com/disneys-pivot-to-streaming-is-a-sign-of-severe-covid-economic-crisis-still-to-come-148360

Anytime a child goes to his or her first movie, it's a memorable experience... Iger, B. (interviewee). (2019). *One Day at Disney*. [Film]. Disney.

Bicycle sales skyrocketed during the COVID-19 pandemic... Abaño, J. (2020). *Bike sales in ANZ skyrocket during COVID-19 lockdown*. Inside Retail. https://insideretail.com.au/news/bike-sales-in-anz-skyrocket-during-covid-19-lockdown-202006

Chapter 8

But the tears are necessary. Huxley, A. (1932). *A Brave New World*. Penguin Random House UK.

A survey in America found that psychologists saw a big jump in the need for their services as a result of the COVID-19 pandemic. Bethune, S. (2021, 9 October). *Demand for mental health treatment continues to increase, say psychologists*. American Psychological Association. https://www.apa.org/news/press/releases/2021/10/mental-health-treatment-demand

In the four weeks to the 19th of September 2021... Australian Institute of Health and Welfare. (2022, February 1). *Mental health services in Australia*. https://www.aihw.gov.au/reports/mental-health-services/mental-health-services-in-australia/report-contents/mental-health-impact-of-covid-19

And according to American psychologist Keith Connors... Conners, C. K. (2000). From the Editor: Attention-deficit/hyperactivity disorder—historical development and overview. *Journal of Attention Disorders, 3*(4), 173–191. https://doi.org/10.1177/108705470000300401

Modern definitions of ADHD don't stray too far from the definitions of Crichton and Still. National Institute of Mental Health. (2021, September). *Attention-Deficit/Hyperactivity Disorder*. National Institute of Mental Health. https://www.nimh.nih.gov/health/topics/attention-deficit-hyperactivity-disorder-adhd

In 1997, 5.5 per cent of children between the ages of three and seventeen were diagnosed with ADHD. Center for Disease Control. (1997). *Summary Health Statistics for U.S. Children: National Health Interview Survey, 1997*. Center for Disease Control. https://www.cdc.gov/nchs/data/series/sr_10/sr10_203.pdf

In 2022, that figure had increased dramatically to 10.2 per cent. Center for Disease Control. (2022). *Interactive Summary Health Statistics for Children*. Center for Disease Control. https://wwwn.cdc.gov/NHISDataQueryTool/SHS_child/index.html

'I struck a match, and I didn't know how much tinder there was around.' Schwarz, A. (2016, December 5). ADHD: the statistics of a "national disaster". *Royal Statistical Society*. https://rss.onlinelibrary.wiley.com/doi/full/10.1111/j.1740-9713.2016.00979.x

Twenty years later in 2022, that number reached a staggering USD $12.42 billion. Insights10. (2023, June 21) US ADHD (Attention Deficit Hyperactivity Disorder) Drugs Market Analysis. https://www.insights10.com/report/us-adhd-attention-deficit-hyperactivity-disorder-drugs-market-analysis/

'the concern is that the pacing of the program, whether it's video games or television, is overstimulating and contributes to attention problems'. Voss, M. (2010). More Screen Time Means More Attention Problems In Kids. *NPR*. https://www.npr.org/sections/health-shots/2010/07/07/128358770/more-screen-time-means-more-attention-problems-in-kids

It's true that if you provide children with a screen device when you go on car trips... Linn, S. (2009). *The Case For Make Believe: Saving Play in a Commercialized World*. New Press.

As Dr Nicholas Kardaras, clinical psychologist and author of Glow Kids... Kardaras, N. (2016). *Glow Kids*. St Martins Griffin

To test whether or not screen time was producing or at least contributing to this behaviour... Dunckley, V. (n.d.). A Four-Week Plan to End Meltdowns, Raise Grades, and Boost Social Skills by Reversing the Effects of Electronic Screen-Time. *Victoria L. Dunckley M.D.* https://drdunckley.com/reset-your-childs-brain/

If ESS occurs in addition to a true underlying psychiatric disorder... Dunckley, V. (2012, July 23). Electronic Screen Syndrome: An Unrecognized Disorder? *Psychology Today.* https://www.psychologytoday.com/au/blog/mental-wealth/201207/electronic-screen-syndrome-unrecognized-disorder

We focus more on that remembered short-term feel-good dopamine surge that we may have experienced in the past... Kardaras, N. (2016). *Glow Kids.* St Martins Griffin.

Postman argued that Huxley, not Orwell, got it right, and this conclusion grows truer each year. Postman, N. (1985). *Amusing Ourselves to Death.* Penguin.

And if ever, by some unlucky chance, anything unpleasant should somehow happen, why, there's always soma to give you a holiday from the facts. This and subsequent quotes - Huxley, A. (1932). *A Brave New World.* Penguin Random House UK.

Chapter 9

One of the greatest dangers we face as we automate the work of our minds... Carr, N. (2011). *The Shallows: What the Internet is Doing to Our Brains.* Norton.

In the second half of 2020 the Netflix documentary *The Social Dilemma* hit screens everywhere. Harris, T. (Interviewee). (2020). *The Social Dilemma.* [Documentary]. Netflix.

www.ingramcontent.com/pod-product-compliance
Lightning Source LLC
Chambersburg PA
CBHW010243010526
44107CB00061B/2666